D1000647

AUTOMATIC MODULATION RECOGNITION OF COMMUNICATION SIGNALS

AUTOMATIC MODULATION RECOGNITION OF COMMUNICATION SIGNALS

by

Elsayed Elsayed Azzouz
Department of Electronic & Electrical Engineering,
Military Technical College,
Cairo, Egypt

and

Asoke Kumar Nandi
Department of Electronic & Electrical Engineering,
University of Strathclyde,
Glasgow, U.K.

KLUWER ACADEMIC PUBLISHERS

BOSTON / DORDRECHT / LONDON

A C.I.P. Catalogue record for this book is available from the Library of Congress

ISBN 0-7923-9796-7

Published by Kluwer Academic Publishers,
P.O. Box 17, 3300 AA Dordrecht, The Netherlands.

Kluwer Academic Publishers incorporates
the publishing programmes of
D. Reidel, Martinus Nijhoff, Dr W. Junk and MTP Press.

Sold and distributed in the U.S.A. and Canada
by Kluwer Academic Publishers,
101 Philip Drive, Norwell, MA 02061, U.S.A.

In all other countries, sold and distributed
by Kluwer Academic Publishers Group,
P.O. Box 322, 3300 AH Dordrecht, The Netherlands.

Printed on acid-free paper

Printed in the Netherlands

To our parents,

Mr. and Mrs. Elsayed Azzouz
and
Dr. and Mrs. Satish Chandra Nandi,

for nurturing our curiosities
and encouraging excellence in us;

and to our families,

Hanan, Bassam, Mohammed, and Islam Azzouz
and
Marion, Robin, David, and Anita Nandi,

for their love, support, and sacrifice.

Contents

Preface

Automatic modulation recognition is a rapidly evolving signal analysis area. In recent years, much interest by academic and military research institutes has focused around the research and development of modulation recognition algorithms. Any communication intelligence (COMINT) system comprises three main blocks: receiver front-end, modulation recogniser and output stage. A considerable work has been done in the area of receiver front-end over the past. The work with the output stage is concerned with information extraction, recording and expoilations, and it is started by signal demodulation, that requires accurate knowledge about the signal modulation type. However, there are two main reasons for knowing the correct modulation type of a signal: to preserve the signal information content and to decide the suitable counter action such as jamming.

The objective of this book is to cover the field of modulation recognition process. The intent is to provide the reader with an understanding of the different techniques. This book aims at an audience consisting of researchers and graduate students as well as practising engineers. In this book we assume that the reader had an introductory course in communication theory and systems, and background in the probability theory and stochastic processes.

The contents of this book have largely been the results of the authors' research over the last few years. A number of very recently published methods for automatic modulation recognition are reviewed critically. This book consists of six chapters and three appendices. Chapter 1 is the introduction. In this chapter, the background and some motivations for the modulation recognition process are introduced. Some classifications of the communication signals that may help in the modulation recognition process are introduced. The problems facing the modulation recognition process are discussed. Furthermore, the mathematical preliminaries required to understand the rest of this book are presented as well as the basic concepts of different analogue and digital modulation types are introduced to help the reader through this book.

The work in this book can be divided into two parts. The first part, which includes Chapters 2 - 4, investigates the use of the decision-theoretic approach for solving the modulation recognition problem. In Chapter 2, a review of five modulation recognisers for analogue modulation recognition is presented. A further five algorithms for analogue modulation recognition - developed by the authors - are introduced. Furthermore, a global procedure for modulation recognition is presented. Computer simulations for different types of analogue modulation corrupted with band-limited Gaussian noise are introduced. Moreover, in this chapter a global procedure for the relevant thresholds determination of the key features necessary for implementing these algorithms is provided. The performance evaluation for one of these recognisers and the overall success rate for all of them are introduced.

A review of ten modulation recognisers for digital modulation recognition is presented in Chapter 3. A further three algorithms for digital modulation recognition - developed by the authors - are introduced. Computer simulations for different types of band-limited digitally modulated signals up to four levels corrupted with band-limited Gaussian noise are introduced. The determination of the relevant thresholds necessary for the implementation of these recognisers is introduced. The performance evaluation for one of these recognisers and the overall success rate for all of them are introduced in some detail.

In Chapter 4, a review of six modulation recognisers for the recognition of both analogue and digital modulations is presented. A further three algorithms for automatic recognition of different types of analogue and digital modulations without any a priori information about the nature of a signal, whether it is analogue or digital - developed by the authors - are introduced. Furthermore, the determination of the relevant thresholds necessary for implementing these algorithms is presented. The performance evaluation for one of these algorithms is introduced.

The second part, which is presented in Chapter 5, investigates the use of the artificial neural networks (ANNs) as another approach for solving the modulation recognition problem. Much work has been done in this book to choose the best ANN structure for modulation recognition. Three types of network structures are considered - no hidden layer, one hidden layer, and two hidden layers. Similar to the decision- theoretic approach, three groups of modulation recognition algorithms, based on the ANN approach, are presented. The first group (one and two hidden layers) is used for the recognition of analogue modulations. The second group (also of one and two hidden layers) is used for the recognition of digital modulations. The third group is used for the recognition of both analogue and digital modulations. Also, a method is introduced for reducing the training time for all these algorithms. Comparisons of the performance evaluations for the algorithms which utilise the decision-theoretic approach with those utilising the ANN approach are provided. Finally, the book is summarised in Chapter 6 and some future directions to extend the work in this book are offered. The references are cited at the end of this book.

This book contains some materials from the authors'research over that last few years. We are grateful to the publishers of our papers for the permission to reproduce some copyright materials in this book.

E. E. Azzouz

A. K. Nandi

September 1996

List of Abbreviations

ADMRAs	*Analogue and digital modulation recognition algorithms.*
AM	*Amplitude modulation.*
AMRAs	*Analogue modulation recognition algorithms.*
ANN	*Artificial neural network.*
ASK2	*Binary amplitude shift keying.*
ASK4	*4-levels amplitude shift keying.*
B.L.	*Band-limited.*
COM.	*Combined modulated signal.*
CW	*Carrier wave (unmodulated signal).*
DMRAs	*Digital modulation recognition algorithms.*
DSB	*Double sideband modulation.*
DT	*Decision-theoretic*
FM	*Frequency modulation.*
FSK2	*Binary frequency shift keying.*
FSK4	*4-levels frequency shift keying.*
FT	*Fourier transform.*
FFT	*Fast Fourier transform.*
IF	*Intermediate frequency.*
IFT	*Inverse Fourier transform.*
IFFT	*Inverse Fast Fourier transform.*
ISB	*Independent sideband modulation.*

LMS	*Least mean square.*
LSB	*Lower sideband modulation.*
MASK	*M-ary amplitude shift keying.*
MFSK	*M-ary frequency shift keying.*
MPSK	*M-ary phase shift keying.*
PBC	*Phase-based classifier.*
PSK2	*Binary phase shift keying.*
PSK4	*4-levels phase shift keying.*
PSK8	*8-levels phase shift keying.*
QLLR	*Quasi log-likelihood ratio.*
RF	*Radio frequency.*
SLC	*Square-law classifier.*
SNR	*Signal-to-noise ratio.*
SSB	*Single sideband.*
Thr.	*Threshold value.*
USB	*Upper sideband.*

List of Symbols

D	Frequency modulation index.
M_s	Number of segments per signal frame.
N_s	Number of samples per segment.
P	Spectrum symmetry measure.
P_{av}	Average probability of correct decision.
Q	Amplitude modulation depth.
R	Chan and Gadbois parameter.
R_{sn}	a coefficient used to determine the amount of noise to be added to a signal at specific SNR.
γ_{max}	Maximum value of the spectral power density of the normalised-centred instantaneous amplitude.
μ_{42}^a	Kurtosis of the normalised-centred instantaneous amplitude.
μ_{42}^f	Kurtosis of the normalised-centred instantaneous frequency.
σ_a	Standard deviation of the instantaneous amplitude in the non-weak intervals of a signal segment.
σ_{aa}	Standard deviation of the absolute value of the normalised-centred instantaneous amplitude.
σ_{af}	Standard deviation of the absolute value of the normalised-centred instantaneous frequency.

σ_{ap}	Standard deviation of the absolute value of the centred non-linear component of the instantaneous phase.
σ_{dp}	Standard deviation of the direct value of the centred non-linear component of the instantaneous phase.
O	Number of output decisions of the ANNs (number of nodes in the output layer).
Qs	Number of training pairs for the ANNs.
Ri	Number of key features (number of nodes in the input layer).
S	Number of nodes in the hidden layer for the single hidden layer ANNs.
$S1$	Number of nodes in the first hidden layer for the two hidden layers ANNs.
$S2$	Number of nodes in the second hidden layer for the two hidden layers ANNs.
SSE	Sum square errors.
$\mathbf{T}$	Actual response of an ANN.
$\mathbf{E}$	Error matrix.
$E\ (i,j)$	The element of $\mathbf{E}$ corresponding to i^{th} row & j^{th} column.

Chapter 1

Introduction

This book is concerned with presenting the most recent algorithms for automatic modulation recognition of communication signals. Some of these algorithms utilise the decision-theoretic approach while the others utilise the ANN approach. In each approach, three groups of modulation recognisers are presented. The first is used for the recognition of analogue modulations only, the second is used for the recognition of digital modulations only, and the third is concerned with both analogue and digital modulations without any a priori information about the nature of a signal. It is found that the developed algorithms are able to distinguish between the different modulation types of interest at a SNR of 10 dB with overall success rate > 97.0% if the modulation types under consideration are analogue modulations only or digital modulations only. Meanwhile, the overall success rate is > 93.0% at a SNR of 15 dB if there is no a priori information about the nature of a signal; i.e. whether it is an analogue or a digitally modulated signal.

In the next section, the problem definition as well as short notes about the manual modulation recognition are introduced. Also, some characteristics of communication signals from the modulation recognition point of view are provided. Also, the importance of knowing the correct modulation type of an incoming RF signal in COMINT applications is considered. Moreover, short notes about the problems facing the modulation recognition process, which may be due to the radio communication channel and the intercept receiver or due to the nature of the intercepted signal, are presented. Meanwhile, more details about the numerical problems associated with the proposed algorithms, which are concerned with the evaluation of the instantaneous amplitude, phase and frequency of a real signal, are presented in Appendix A.

1.1 Background and Motivations

Communication signals travelling in space with different modulation types and different frequencies fall in a very wide band. Usually, it is required to identify and monitor these signals for many applications. Some of these applications are for civilian purposes such as signal confirmation, interference identification and spectrum management. Civilian authorities may wish to monitor their transmissions in order to maintain a control over these activities as well as detecting and monitoring the non-licensed transmitters. The other applications are for military purposes such as electronic warfare (EW), surveillance and threat analysis. In electronic warfare applications, electronic support measures (ESM) system plays an important role as a source of information required to conduct electronic counter measures (ECM), electronic counter-counter measures (ECCM), threat detection, warning, target acquisition and homing.

In the past, COMINT systems have relied on the operator interpretation (manual modulation recognition) of measured parameters to provide classification of different emitters. In manual modulation recognition, the search equipment configuration as shown in Fig. 1.1 makes available for the search operator the following four types of information: *1) IF time waveform, 2) spectrum of a signal (instantaneous and averaged), 3) instantaneous amplitude (AM detector output) and instantaneous frequency (FM detector output) of the intercepted signal, 4) sound of a signal.* The importance of these information is explained as below:

1. The IF time waveform is useful for studying the time characteristics of a signal. Also, it can help in determining if there is transmission or not. It can be displayed on an oscilloscope.

2. The spectral analysis is useful for studying the signal spectra at the the IF frequency. During the search mode, the receiver IF bandwidth should be made as wide as possible, to see all the activities in the frequency band of interest on the spectrum analyser display. After that, only one signal is chosen for modulation recognition and further analysis. Spectral analysis helps in determining the activities as well as in multi-channel signal analysis. Furthermore, it helps to

determine if the signal has a carrier component or not. Moreover, it can be used as templates for the hostile transmitters.

3. The AM and FM demodulator displays are used for displaying the amplitude and frequency modulation content in a signal. AM display indicates the change of amplitude of a carrier, thus, it can be used for AM and MASK analysis. FM display represents the change in the frequency of a signal, thus it can be used for FM, MFSK and MPSK analysis. Finally, the AM and FM demodulators outputs are displayed on a dual beam oscilloscope.

4. The audio analysis is performed by listening to the head phones connected to the demodulators. Audio recognition is powerful because the operator can quickly detect the cyclic rate in the signal under consideration such as pitch changes. Also, the sound of the head phone output helps in adjusting the oscilloscope's time base.

Recently, several modulation recognisers (identifiers or classifiers) which automatically recognise the type of modulation present in an incoming RF signal have been developed. One of the oldest version of modulation recognisers [5] uses a bank of demodulators, each designed for only one type of modulation. An operator examining or listening to the demodulator outputs could decide about the modulation type of the received signal. This recogniser however requires long signal durations and highly skilled operators. The automation of this recogniser is achieved by introducing a set of intelligence decision algorithms at the demodulator outputs as shown in Fig. 1.2. However, the implementation of this recogniser is complex and requires excessive computer storage. Moreover, the number of modulation types that can be recognised is based on the number of demodulators used.

Automatic modulation recognition is more powerful than manual modulation recognition because by integrating the automatic modulation recogniser into an electronic support measurement (ESM) receiver, including an energy detector and a direction finder (DF), would allow an operator to improve his efficiency and his ability to monitor the different activities in the frequency band of interest. So, in advanced ESM systems, the operator is replaced by sophisticated electronic machines. The function of these machines is to exploit the electro-magnetic emissions of the other side for the purpose of automatic gathering intelligence information. The main objective of any surveillance system is threat recognition by comparing the characteristics of the intercepted emitters against a catalogue of reference characteristics or signal sorting parameters. One of the important parameters is the signal modulation type.

Generally, any surveillance system in COMINT applications consists of three main blocks: receiver-front-ends (activity detection and frequency down conversion), modulation recogniser (key features extraction and classification), and output stage (normal demodulation and information extraction). There are several types of receiver-front ends such as channelised, acousto-optical spectrum analyser, instantaneous frequency measurement (IFM), scanning superheterodyne, and micro-scan superheterodyne receivers and there are several references dealing with this part (e.g. [1]-[3], and the references there in). At the output stage there are several functions performed and they are mainly related to information extraction, recording and exploitations. All these functions are preceded by signal demodulation. Once the correct modulation type of the intercepted signal is determined, all the following functions performed at the output stage are straightforward classical functions, but may not be trivial (e.g. deciphering). So, the key functional block is the modulation recogniser. The prior information required for any modulation recogniser is the signal bandwidth, which can be determined through the use of an energy detector in the receiver front-end stage. The information obtained by the energy detector, modulation recogniser, and parameters estimator such as the carrier frequency, the signal bandwidth, the bit rate, the modulation type, ... etc. are gathered to perform the signal demodulation and information extraction.

Modulation recognition is extremely important in COMINT applications for several reasons. Firstly, applying the signal to an improper demodulator may partially or completely damage the signal information content. Secondly, knowing the correct modulation type helps to recognise the threat and to determine the suitable jamming waveform.

In the following three chapters, a review of the more recent papers published in the area of modulation recognition during 1984 - 1995 [4]-[24], as the first paper in this area [4] have been published, is introduced. Because of the classified nature of the problem, many details are not mentioned in the available references. The available references can be classified into three categories according to the modulation types for each of them. The first category is concerned with the recognition of analogue modulations only [7], [10], [15], [19], and [23]. The second category is concerned with the recognition of digital modulations only [4], [9], [13], [14], [17], [18], [20], [21], [22], and [24]. The third category is concerned with the recognition of both analogue and digital modulations without any a priori information [5], [6], [8], [11], [12], and [16]. It is worth noting that in the first category, which is concerned with the recognition of analogue modulations only, none of the available references considered VSB, and the combined modulations. Also, in the last category, which is concerned with the recognition of both analogue and digital modulations, not all the well known modulation types that will be considered throughout the book, were considered in the available references but only subset of these types. On the other hand, some of the modulation recognisers presented in the available references were tested by real signals such as [6], [11], and [12].

Generally, there are two philosophies for approaching the modulation recognition problems in the available references namely, 1) a decision-theoretic approach and 2) a statistical pattern recognition approach. In the decision-theoretic approach, probabilistic and hypotheses testing arguments are employed to formulate the modulation recognition problem. In the statistical pattern recognition approach, the classification system is divided into two subsystems. The first is a features extraction subsystem, whose function is to extract the pre-defined features from the received data; i.e reducing the dimensions of the pattern representation. The second subsystem is a pattern recognition subsystem, whose function is to indicate the modulation type of a signal. The work with pattern recognition comprises two phases - 1) a training phase to adjust the classifier structure and 2) a test phase that gives the classification decision. The modulation recognisers, in the available references, were developed according to either approach. There are also some recognisers combining the two approaches.

In this book, another approach to the modulation recognition process, which utilise the ANNs, is developed. This approach comprises three main steps: 1) pre-processing and key features extraction, 2) a training stage to adjust the classifier structure (determine the weights and biases of the chosen networks), and 3) a test phase, in which the performance evaluation of the chosen network is determined. Both the decision-theoretic and the ANN algorithms introduced in this book are characterised by the simplicity of key features extraction and the high speed of computations. Furthermore, less computer storage is required for these approaches than the pattern recognition approach. These motivate the use of the presented decision-theoretic and ANN algorithms in the real-time analysis.

Also, there are five techniques for solving the modulation recognition problem. These are: 1) spectral processing, 2) instantaneous amplitude, phase and frequency parameters, 3) instantaneous amplitude, phase and frequency histograms, 4) combinations of the previous three methods and 5) universal demodulators. The work in this book utilise mainly the second technique. The reasons for choosing this technique are: it is simple, requires less memory and is faster than the other techniques. So, the developed recognisers are likely to be applicable in the real field and to be suitable for on-line analysis.

From the modulation recognition point of view, there are many classifications for communication signals as shown in Fig. 1.3. The first classification is according to the signal information content. In this classification any communication signal can be categorised as one of four categories

1. If the signal has no phase information, it is called amplitude signal. In this case the instantaneous phase is a linear function of time; $\phi(t) = At + B$ where A and B are constants, i.e. the amplitude signal has amplitude information and no phase information.

2. If the signal has no amplitude information, it is called phase signal and its instantaneous amplitude is constant; $a(t) = C$, where C is constant, i.e. the phase signal has phase information and no amplitude information.

3. If the signal has both amplitude and phase information, it is called Combined signal; the useful information is in both the instantaneous amplitude and phase or frequency of a signal.

It is worth noting that, if the signal has no phase information and no amplitude information, it is called CW signal. In this case the instantaneous phase is a linear function of time and the instantaneous amplitude is constant, i.e. the CW signal has no useful information (no amplitude information and no phase information).

The second classification is according to the signal spectrum symmetry around its carrier frequency. Generally, the RF signal spectrum is composed of a carrier component plus two side-bands but in some types one or two of these components may be lost or reduced. Based on this fact for example, there are many types of amplitude modulated signals such as: AM, DSB, LSB, USB, VSB, ISB (see Fig. 1.4). Thus, due to the existence of the side-bands, any communication signal can be categorised as in one of two categories: 1) symmetric signals and 2) asymmetric signals. A perfectly symmetric signal is characterised by the signal power in the two side-bands being the same. An asymmetric signal is characterised by one of the two side-bands being reduced or disappearing and so the signal power in the two side-bands is very different. The third classification is according to the nature of the modulating signal used. In this classification, any communication signal can be categorised as in one of two categories: 1) analogue modulated signal in which the modulating signal is an analogue signal (a continuous time - continuous amplitude signal) such as speech signal, video signal,..., and 2) digitally modulated signal, in which the modulating signal is a digital symbol sequence (a continuous time - discrete amplitude signal). However, more details about the different types of analogue and digital modulations are introduced in Section 1.3.

In the first classification, it is required for any modulation recogniser to determine where the intercepted signal contains the useful information; i.e. to identify whether the information exists in the instantaneous amplitude, the instantaneous phase, the instantaneous frequency, or a combination of them. In the second classification, it is necessary to measure the signal spectrum symmetry around its carrier frequency. In the third classification, it is necessary to know the source of the modulating signal. The work in this book uses these three classifications. It is worth noting that the first classification is used in all the modulation recognition algorithms presented in this book. The second classification is not required for the recognition algorithms of digital modulations but it is required for the other algorithms. The third classification is only required in the proposed recognition algorithms for both analogue and digital modulations (no a priori information about the nature of a signal).

Modulation recognition brings together many aspects of cooperative communication theory such as signal detection, parameter estimation, channel identification and tracking. Furthermore, modulation recognition environments may vary between two extremes - from no significant noise in the best situations to a very noisy one with inter-ference and fading. Moreover, there are many practical problems facing the modulation recognition process. Some of these problems are due to the radio communication chan-nel and the intercept receiver. The other problems are due to the nature of the received signal. The problems arising from the radio communication channel and the intercept receiver should be solved in the receiver front-ends (pre- processing) stage for perfect modulation recognition.

The problems such as multi-path fading effects, weak signals reception, signal distor-tion, frequency instability, interference from adjacent channels and signal selection are due to the radio communication channel and the intercept receiver. It is worth noting that the presence of fading introduces an undesirable envelope modulation which can confuse the envelope features of the received signals. To avoid the fading-modulation effects, the envelope should be segmented; each segment length depends on the fading bandwidth. But, the choice of the segment length must be long enough to allow good features extraction from each segment. Also, it is necessary to isolate only one of the activities in the COMINT receiver bandwidth to decide about its modulation type. The relevant information for this isolation are derived from the spectral estimation of the received signal and it should be done in the pre-processing stage. Also, there are some problems related to the signal nature such as the weak segments of a signal (carrier absent or reduced and the pauses in transmission in analogue modulations), lower SNR reception, and the transmission time and the speed of computations. Throughout the book, some of these problems are discussed and the suitable solutions will be presented in the appropriate places.

Most of the key features used in the modulation recognition algorithms in this book are derived from three important parameters. These are the instantaneous amplitude, the instantaneous phase and the instantaneous frequency. There are some numerical problems associated with the evaluation of these parameters such as: choice of sam-pling rate, weak signal intervals, phase wrapping, linear phase component and the numerical derivative. The details of these problems and suitable solutions for them are introduced in Appendix A. Also, in these modulation recognition algorithms, the carrier frequency needs to be known as some of the key features used are derived from the instantaneous phase and the instantaneous frequency of a signal. Furthermore, the spectrum symmetry measure, that is used as a key feature in some of the developed algorithms, requires the exact value of the carrier frequency to be known. So, a new method for very good carrier frequency estimation, based on the zero-crossing sequence in the non-weak intervals of a signal segment only, as well as comparisons with other two methods, are introduced in Appendix B.

1.2   Mathematical Preliminaries

The digital processing of broad band signals requires high sampling rate and hence both the memory size and the processing speed must be increased. All the processing of the received data vector should be completed before the arrival of a new data sample and construction of a new data vector. In practice, we seek a minimal representation of the data; i.e minimum frequency bandwidth in order to lower the sampling rate. All real signals have redundancy (unnecessary repetition), which can be removed. If $x(t)$ is real, then from the hermitian symmetry $X(f) = X^*(-f)$; i.e. whole signal information content can be had from only one half of the signal spectrum. So, we can represent any real signal by its right half spectrum. This representation is called *the analytic representation*. The digital processing of the analytic representation requires half the sampling rate required for the broad band real signal, but the same memory size because the derived samples are complex.

The Hilbert transform, $y(t)$, of a real signal, $x(t)$, is the result of applying $x(t)$ to a quadrature filter, F_Q, with impulse response $r_Q(t)$ and complex gain $G_Q(f)$ [25]. Thus, we can write

$$y(t) = F_Q\{x(t)\} \tag{1.1}$$
$$= x(t) * r_Q(t)$$
$$= \int_{-\infty}^{\infty} x(t-\theta).r_Q(\theta)d\theta, \tag{1.2}$$

where,

$$r_Q(t) = \frac{1}{\pi t} \tag{1.3}$$

Substitution from (1.3) in (1.2) the Hilbert transform can be expressed as

$$y(t) = P.V. \int_{-\infty}^{\infty} \frac{x(t-\theta)}{\pi\theta}d\theta$$

$$= P.V. \int_{-\infty}^{\infty} \frac{x(\theta)}{\pi(t-\theta)} d\theta \tag{1.4}$$

where P.V. denotes the principle value of the integral. The complex gain of the quadrature filter is given by

$$G_Q(f) = \frac{Y(f)}{X(f)} = -j\,signum(f) \tag{1.5}$$

Also, any real signal can be represented by its right half spectrum through the use of the the analytic representation of a signal. The analytic signal, $z(t)$, associated with a real signal, $x(t)$, is obtained by applying the real signal to an analytizing filter, F_A. This filter can be decomposed into an identity filter, F_I, plus a quadrature filter, F_Q, such that:

$$F_A = F_I + jF_Q. \tag{1.6}$$

The identity filter is characterised by an impulse response $r_I(t) = \delta(t)$ and a complex gain $G_I(f) = 1$, while the impulse response and the complex gain of the quadrature filter are given by (1.3) and (1.5) respectively. Thus the analytic signal, $z(t)$, can be expressed as

$$
\begin{aligned}
z(t) &= F_A\{x(t)\} \\
&= [F_I + jF_Q]\{x(t)\} \\
&= x(t) + jy(t) \tag{1.7}
\end{aligned}
$$

Thus the analytic signal, $z(t)$, is a complex function, its real part is the real signal, $x(t)$, and its imaginary part, $y(t)$, is the Hilbert transform of $x(t)$. The spectrum of the analytic signal, $Z(f)$ is given by

$$
\begin{aligned}
Z(f) &= X(f) + jY(f) \\
&= [1 + signum(f)]\,X(f). \\
&= 2U(f)X(f) \tag{1.8}
\end{aligned}
$$

where $U(f)$ is the unit step in frequency domain and it is defined as

$$
U(f) = \begin{cases} 1 & \text{if} \quad f > 0 \\ \frac{1}{2} & \text{if} \quad f = 0 \\ 0 & otherwise \end{cases} \tag{1.9}
$$

The complex envelope, $\alpha(t)$, of a real signal, $x(t)$, is derived from its analytic representation as indicated in (1.10) below

$$\alpha(t) = z(t) \; e^{-j2\pi f_c t}, \tag{1.10}$$

where f_c is some arbitrary frequency. For example, in case of narrow band signals f_c is usually taken as the carrier frequency. From (1.7) and (1.10), $\alpha(t)$ can be expressed as [26]

$$\alpha(t) = m(t) + jn(t), \tag{1.11}$$

where

$$m(t) = x(t)\cos(2\pi f_c t) + y(t)\sin(2\pi f_c t), \tag{1.12}$$

and

$$n(t) = y(t)\cos(2\pi f_c t) - x(t)\sin(2\pi f_c t). \tag{1.13}$$

The practical implementation of (1.12) and (1.13) is shown in Fig. 1.5a. It is worth noting that the complex envelope may be pure real, pure imaginary or complex function, but the analytic representation is always complex function. Also, $x(t)$ can be reconstructed from $m(t)$ and $n(t)$ using the following analytic form

$$x(t) = m(t)\cos(2\pi f_c t) - y(t)\sin(2\pi f_c t) \tag{1.14}$$

The practical implementation of (1.14) is shown in Fig. 1.5b.

The instantaneous amplitude and the instantaneous phase of a signal can be evaluated either from the analytic representation of a signal using (1.7) or from the complex envelope representation using (1.11). The instantaneous amplitude, $a(t)$, is defined as

$$\begin{aligned} a(t) & = \mid z(t) \mid = \sqrt{x^2(t) + y^2(t)} \\ & = \mid \alpha(t) \mid = \sqrt{m^2(t) + n^2(t)}. \end{aligned} \tag{1.15}$$

The instantaneous phase, $\phi(t)$, can be calculated from the analytic representation as

$$\phi(t) = \begin{cases} tan^{-1}\left[\frac{y(t)}{x(t)}\right] & \text{if } x(t) > 0, y(t) > 0 \\ \pi - tan^{-1}\left[\frac{y(t)}{x(t)}\right] & \text{if } x(t) < 0, y(t) > 0 \\ \frac{\pi}{2} & \text{if } x(t) = 0, y(t) > 0 \\ \pi + tan^{-1}\left[\frac{y(t)}{x(t)}\right] & \text{if } x(t) < 0, y(t) < 0 \\ \frac{3\pi}{2} & \text{if } x(t) = 0, y(t) < 0 \\ 2\pi - tan^{-1}\left[\frac{y(t)}{x(t)}\right] & \text{if } x(t) > 0, y(t) < 0 \end{cases} \tag{1.16}$$

Similarly, $\phi(t)$ can be calculated from the complex envelope. The only difference between the two ways of calculation is the linear phase component due to the carrier component; i.e. $\phi(t) = \arg\{z(t)\} = \arg\{\alpha(t)\} + 2\pi f_c t$. The practical implementation of (1.15) and (1.17) is shown in Fig. 1.6a and Fig. 1.6b respectively. The division operation in (1.16) is implemented in practice as shown in Fig. 1.6c.

Finally the instantaneous frequency, $f(t)$, is given by

$$f(t) = \frac{1}{2\pi} \frac{d\phi(t)}{dt}. \tag{1.17}$$

1.3 General Concepts about Modulation Techniques

In this section, the analytic expressions of the instantaneous amplitude, $a(t)$, phase, $\phi(t)$, and frequency, $f(t)$, as well as the spectral power density of different types of modulation are introduced. These parameters are important features of any real RF signal as they represent most of the features in a signal. Furthermore the graphical representation of these parameters for each modulation type is introduced. For the graphical representation, the carrier frequency, f_c, and the sampling rate, f_s, were respectively chosen to 150 kHz and 1200 kHz. A non-intelligible speech signal with cut-off frequency ought to be 8 kHz is used as a modulating signal for analogue modulations. Also, a random symbol sequence at symbol rate equal to 12.5 kHz is used as a modulating signal for digital modulations. The procedure for generating these modulating signals will be explained in Chapters 2 and 3.

1.3.1 Analogue modulated signals

The time-domain and frequency-domain description (both mathematically and graphically) for different analogue modulations are introduced . Furthermore, the mathematical expressions for the instantaneous amplitude and phase of analogue modulated signal are presented.

1. Amplitude modulation (AM)

For AM signal, the modulated signal, $s(t)$, can be expressed as [27]

$$s(t) = [1 + mx(t)] \cos(2\pi f_c t), \tag{1.18}$$

where, m is the required amplitude modulation depth, $x(t)$ is the modulating (message) signal, and f_c is the carrier frequency. Its Fourier transform is given by

$$\begin{aligned} S(f) &= (1/2)\{\delta(f - f_c) + \delta(f + f_c)\} \\ &+ (m/2)\{X(f - f_c) + X(f + f_c)\} \end{aligned} \tag{1.19}$$

Where $X(f)$ is the Fourier transform of the modulating signal, $x(t)$, and $\delta(f)$ is the Dirac delta function. From (1.19) and Fig. 1.7, it is clear that $S(f)$ is a frequency translated version of the modulating signal plus a carrier component. Furthermore, the transmission bandwidth is $B = 2W$, where W is the modulating signal bandwidth.

By straightforward analysis, the analytic signal, $z(t)$, is given by

$$z(t) = [1 + mx(t)] e^{j2\pi f_c t} \tag{1.20}$$

Hence, the instantaneous amplitude, $a(t)$, and the the instantaneous phase, $\phi(t)$, can be expressed as

$$a(t) = |1 + mx(t)|, \tag{1.21}$$

and

$$\phi(t) = 2\pi f_c t \tag{1.22}$$

From (1.21), (1.22) and Fig. 1.7, the instantaneous amplitude is time varying function while the instantaneous phase after removing the linear phase component is constant ($= \mathbf{0}$).

Example (1)

In the case of single harmonic modulating signal, $s(t)$ is expressed as

$$s(t) = [1 + Q\cos(2\pi f_x t)]\cos(2\pi f_c t), \tag{1.23}$$

where f_x is the modulating signal frequency and Q is the required modulation depth. Its Fourier transform, $S(f)$, is given by

$$
\begin{aligned}
S(f) &= (1/2)\{\delta(f - f_c) + \delta(f + f_c)\}\\
&+ (Q/4)\{\delta[f - (f_c + f_x)] + \delta[f + (f_c + f_x)]\}\\
&+ (Q/4)\{\delta[f - (f_c - f_x)] + \delta[f + (f_c - f_x)]\}
\end{aligned} \tag{1.24}
$$

Similarly, by straightforward analysis, the analytic signal, $z(t)$, is given by

$$z(t) = [1 + Q\cos(2\pi f_x t)]\, e^{j2\pi f_c t} \tag{1.25}$$

Hence, the instantaneous amplitude, $a(t)$, can be expressed as

$$a(t) = |\, 1 + Q\cos(2\pi f_x t)\,|, \tag{1.26}$$

and the instantaneous phase, $\phi(t)$, is given by

$$\phi(t) = 2\pi f_c t \tag{1.27}$$

2. Double side-band modulation (DSB)

For DSB signal, $s(t)$, is expressed as [27]

$$s(t) = x(t)\cos(2\pi f_c t), \tag{1.28}$$

Its Fourier transform is given by

$$S(f) = (1/2)\{X(f - f_c) + X(f + f_c)\} \tag{1.29}$$

From (1.29) and Fig. 1.8, it is clear that $S(f)$ is a frequency translated version of the modulating signal only (no carrier component). Also, the transmission bandwidth is $B = 2W$.

Similarly, the analytic signal associated with $s(t)$ is given by

$$z(t) = x(t)e^{j2\pi f_c t} \qquad (1.30)$$

So, the instantaneous amplitude and phase are given by

$$a(t) = |x(t)|, \qquad (1.31)$$

and

$$\phi(t) = \begin{cases} 2\pi f_c t & \text{if } x(t) > 0 \\ 2\pi f_c t + \pi & \text{if } x(t) < 0 \end{cases} \qquad (1.32)$$

From (1.31), (1.32) and Fig. 1.8, the instantaneous amplitude is time varying function while the instantaneous phase after centring and removing the linear phase component takes only two values (- $\pi/2$ and $\pi/2$).

Example (2)

In case of single harmonic modulating signal, $s(t)$ is expressed as

$$s(t) == \cos(2\pi f_x t)\cos(2\pi f_c t) \qquad (1.33)$$

Its Fourier transform is given by

$$\begin{aligned} S(f) &= (1/4)\{\delta[f - (f_c + f_x)] + \delta[f + (f_c + f_x)]\} \\ &+ (1/4)\{\delta[f - (f_c - f_x)] + \delta[f + (f_c - f_x)]\} \end{aligned} \qquad (1.34)$$

Similarly, the analytic signal is given by

$$z(t) = \cos(2\pi f_x t)e^{j2\pi f_c t} \qquad (1.35)$$

So, the instantaneous amplitude and phase are given by

$$a(t) = |\cos(2\pi f_x t)|, \qquad (1.36)$$

and

$$\phi(t) = \begin{cases} 2\pi f_c t & \text{if } \cos(2\pi f_x t) > 0 \\ 2\pi f_c t + \pi & \text{if } \cos(2\pi f_x t) < 0 \end{cases} \qquad (1.37)$$

3. Single side-band modulation (SSB)

The AM and the DSB modulations require a transmission bandwidth twice the message bandwidth. In either case, one half of the transmission bandwidth is occupied by the upper side-band (USB), while the other half is occupied by the lower side-band (LSB). However, the USB and the LSB contain the same information. This means only one side-band is necessary to transmit the required information. In this case, the modulated signal bandwidth is equal to the modulating signal bandwidth ($B = W$).

For SSB signal, $s(t)$ is expressed as [27]

$$s(t) = x(t)\cos(2\pi f_c t) \mp y(t)\sin(2\pi f_c t), \tag{1.38}$$

where $x(t)$ is the modulating signal and $y(t)$ is its Hilbert transform. The upper sign is used for USB signal generation and the lower sign is used for the LSB signal generation. Its Fourier transform is given by

$$
\begin{aligned}
S(f) &= (1/2)\{X(f - f_c) \pm jY(f - f_c)\} \\
&+ (1/2)\{X(f + f_c) \mp jY(f + f_c)\}
\end{aligned}
\tag{1.39}
$$

But,

$$
Y(f) = \begin{cases} -j.X(f) & \text{if } f > 0 \\ j.X(f) & \text{if } f < 0 \end{cases}
\tag{1.40}
$$

Thus,

$$
\begin{aligned}
S(f) = (1/2) &\begin{cases} X(f - f_c) \pm X(f - f_c) & \text{if } (f - f_c) > 0 \\ X(f - f_c) \mp X(f - f_c) & \text{if } (f - f_c) < 0 \end{cases} \\
+ (1/2) &\begin{cases} X(f + f_c) \mp X(f + f_c) & \text{if } (f + f_c) > 0 \\ X(f + f_c) \pm X(f + f_c) & \text{if } (f + f_c) < 0 \end{cases}
\end{aligned}
\tag{1.41}
$$

From (1.41) and Figs. 1.9, and 1.10, it is clear that $S(f)$ for USB signal exists for $f > f_c$ (150 kHz) and for LSB, $S(f)$ exists $f < f_c$ (150 kHz).

For example, the Fourier transform of USB signal is given by

$$S(f) = \begin{cases} X(f - f_c) & \text{if } f > f_c \\ X(f + f_c) & \text{if } f < -f_c \end{cases} \tag{1.42}$$

Let $x(t)$ be the sum of N harmonics; i.e.

$$x(t) = \sum_{i=1}^{N} x_i \cos(2\pi f_i t + \psi_i), \quad f_N < f_x, \tag{1.43}$$

So $y(t)$, which is the Hilbert transform of $x(t)$, is expressed as

$$y(t) = \sum_{i=1}^{N} x_i \sin(2\pi f_i t + \psi_i) \tag{1.44}$$

Thus, the SSB signal can be expanded as

$$s(t) = \sum_{i=1}^{N} x_i \cos\left(2\pi(f_c \pm f_i)t + \psi_i\right) \tag{1.45}$$

Also, the Hilbert transform $y_{SSB}(t)$ of $s(t)$ can be written as

$$y_{SSB}(t) = \sum_{i=1}^{N} x_i \sin\left[2\pi(f_c \pm f_i)t + \psi_i\right] \tag{1.46}$$

The instantaneous amplitude and phase are given by

$$a(t) = \sqrt{\sum_{i=1}^{N} x_i^2 + 2\sum_{i=1}^{N}\sum_{j=1}^{N} x_i x_j \cos\left[2\pi(f_i - f_j)t\right]}, \tag{1.47}$$

and

$$\phi(t) = \tan^{-1} \frac{\sum_{i=1}^{N} x_i \sin\left[2\pi(f_c + f_i)t + \psi_i\right]}{\sum_{i=1}^{N} x_i \cos\left[2\pi(f_c + f_i)t + \psi_i\right]} \tag{1.48}$$

From (1.47), (1.48) Fig. 1.9 and Fig. 1.10, it is clear that both $a(t)$ and $\phi(t)$ are time varying functions.

4. Vestigial side-band modulation (VSB)

SSB modulation is the most suitable type if there is an energy gap between the zero and the lowest frequency component of the modulation signal spectrum. But some modulating signals contain significant components at extremely low frequencies (TV signals and wide band data). In this case the SSB is inappropriate. The VSB is used to overcome this difficulty. In this case, the transmission bandwidth $W < B < 2W$.

Generally, the VSB signal is derived from AM signals by filtering through a vestigial side band filter [27]. The details of the magnitude and phase response of the VSB filter will introduced in Chapter 2. The Fourier transform of the VSB modulated signal is given by,

$$S(f) = S_{AM}(f).H(f) \tag{1.49}$$

where, $S_{AM}(f)$ is the spectrum of AM signal and $H(f)$ is the complex gain of the VSB filter. Thus,

$$
\begin{aligned}
S(f) &= (1/2)\left\{X(f - f_c) + \delta(f - f_c)\right\} H(f) \\
&+ (1/2)\left\{X(f + f_c) + \delta(f + f_c)\right\} H(f)
\end{aligned}
\tag{1.50}
$$

As the VSB is derived from AM signal by filtering, its spectrum is an intermediate between the AM and the SSB signals as shown in Fig. 1.11.

5. Angle modulated signals

For angle modulated signals, either the phase or frequency of the carrier is varied according to the modulating signal. Generally, the angle modulated signal can be expressed as

$$s(t) = \cos\left[\theta(t)\right] \tag{1.51}$$

where $\theta(t)$ is varied as the modulating signal, $x(t)$, varied. There are infinite number of methods in which $\theta(t)$ may be varied in some manner with the modulating signal. Only two methods are considered in this chapter.

a) Frequency modulation (FM): in which the instantaneous frequency is varied linearly with the modulating signal $x(t)$ and it is expressed by [27]

$$s(t) = \cos\left[2\pi f_c t + K_f \int_{-\infty}^{t} x(\tau)d\tau\right], \tag{1.52}$$

where K_f is the frequency deviation coefficient and it will be defined in Chapter 2.

b) Phase modulation (PM): in which the instantaneous phase is varied linearly with the modulating signal $x(t)$ and it is expressed by

$$s(t) = \cos\left[2\pi f_c t + K_p x(t)dt\right], \tag{1.53}$$

where K_p is the phase deviation coefficient.

Let us consider the case of frequency modulation, its Fourier transform is given by

$$S(f) = (1/2)\left\{G(f - f_c) + G^*(f + f_c)\right\} \tag{1.54}$$

where $G(f)$ is the Fourier transform of $\left\{e^{jK_f \int_{-\infty}^{t} x(\tau)d\tau}\right\}$. In case of single harmonic modulating signal,

$$G(f) = \sum_{n=-\infty}^{n=\infty} J_n(\beta)\delta(f - nf_x) \tag{1.55}$$

where $J_n(\beta)$ is the nth order Bessel function. Since the FM waveform has a constant power level, so through the traffic the side bands appear and the component at the carrier frequency decreases as shown in Fig. 1.12. The variation in the carrier and side bands components level is based on the value of the modulation index. At certain values of the modulation index (1.405, 5.520, 8.6, ...) the component at the carrier frequency disappear. Also, at other values (0, 3.8, 10.2, ...) the first side-band component disappears and at some other values (0, 5, 8.1, ...) the second side-band component disappears.

The Hilbert transform, $y(t)$, of the modulated signal, $s(t)$, is given by

$$y(t) = \sum_{-\infty}^{\infty} J_n(\beta) \sin \mid 2\pi f_c t + 2\pi n f_x t \mid \qquad (1.56)$$

So, the instantaneous amplitude and phase are given by

$$a(t) = 1, \qquad (1.57)$$

and

$$\phi(t) = \tan^{-1} \frac{\sum_{-\infty}^{\infty} J_n(\beta) \sin \mid 2\pi (f_c + n f_x) t \mid}{\sum_{-\infty}^{\infty} J_n(\beta) \cos [2\pi (f_c + n f_x) t]} \qquad (1.58)$$

From (1.57), (1.58) and Fig. 1.12, $a(t)$ is constant and $\phi(t)$ is time varying function.

6. Combined modulated signals

In this type two modulating signals transmitted at the same carrier frequency, the first is amplitude modulated signal and the second is frequency modulated signal. So the combined modulated signal can be expressed as

$$s(t) = [1 + Q x_1(t)] \cos \left[2\pi f_c t + 2\pi K_f \int_{-\infty}^{t} x_2(\tau) d\tau \right], \qquad (1.59)$$

where $x_1(t)$ and $x_2(t)$ are two the modulating signals - $x_1(t)$ is used for amplitude modulation and $x_2(t)$ is used for frequency modulation. For more details about combined modulation see Chapter 2 and Fig. 1.13. It is clear that both $a(t)$ and $\phi(t)$ are time varying functions.

1.3.2   Digitally modulated signals

1. Amplitude shift keying (ASK)

For binary ASK signal, $s(t)$, is represented by [28]

$$s(t) = m(t) \cos(2\pi f_c t) \qquad (1.60)$$

where

$$m(t) = \begin{cases} 1 & 0 \leq t \leq T_b \\ 0 & 0 \leq t \leq T_b \end{cases} \qquad (1.61)$$

and T_b is the bit duration $(= 1/r_b)$, r_b is the bit rate. The spectral power density $G_s(f)$ is given by [28]

$$
\begin{aligned}
G_s(f) &= \frac{A^2}{16}\left[\delta(f - f_c) + \delta(f + f_c)\right] \\
&+ \frac{A^2}{16}\left[\frac{\sin^2 \pi T_b(f - f_c)}{\pi^2 T_b(f - f_c)^2} + \frac{\sin^2 \pi T_b(f + f_c)}{\pi^2 T_b(f + f_c)^2}\right]
\end{aligned}
\tag{1.62}
$$

From (1.62) and Fig. 1.14, it is clear that the spectrum of the ASK2 signal comprises of $(Sinc)^2$ function plus a carrier component. By straight forward analysis, the complex envelope is given by

$$
\alpha(t) = m(t) \tag{1.63}
$$

The instantaneous amplitude can be expressed as

$$
\begin{aligned}
a(t) &= |m(t)| \\
&= \begin{cases} 0 & \text{if m(t) = 0} \\ 1 & \text{if m(t) = 1} \end{cases}
\end{aligned}
\tag{1.64}
$$

and the instantaneous phase is given by

$$
\phi(t) = 0 \tag{1.65}
$$

From (1.64), (1.65) and Fig. 1.14, the instantaneous amplitude looks like the bit stream, while the instantaneous phase is always 0. For 4-levels ASK, see Fig. 1.15.

2. Phase shift keying (PSK)

The PSK2 signal is represented by [28]

$$
s(t) = \cos\left(2\pi f_c t + D_p m(t)\right) \tag{1.66}
$$

where D_p is the modulation index of PSK signal and $m(t) = \pm 1$ instead of (0,1) in ASK2. The spectral power density $G_s(f)$ is given in [28]

$$
G_s(f) = \frac{A^2}{4}\left[\frac{\sin^2 \pi T_b(f - f_c)}{\pi^2 T_b(f - f_c)^2} + \frac{\sin^2 \pi T_b(f + f_c)}{\pi^2 T_b(f + f_c)^2}\right]. \tag{1.67}
$$

From (1.67) and Fig. 1.16, the spectrum of PSK signal comprises of a $Sinc^2$ function and no carrier component. Also, by straight forward analysis and by letting $D_p = \pi/2$, which achieve the maximum power in the signal [28], $s(t)$ can be re-expressed as

$$s(t) = -m(t)\sin(2\pi f_c t) \tag{1.68}$$

Thus, the complex envelope is given by

$$\alpha(t) = j.m(t) \tag{1.69}$$

Also, the instantaneous amplitude and phase are given by

$$a(t) = \mid m(t) \mid = 1 \tag{1.70}$$

$$\phi(t) = \begin{cases} -\pi/2 & \text{if m(t) = -1} \\ \pi/2 & \text{if m(t) = 1} \end{cases} \tag{1.71}$$

From (1.70) the instantaneous amplitude should be constant but due to the band-limitation that is used in the computer simulations of different modulation types as will explain in details in Chapters 2 and 3, there is a minor amplitude variations especially at the transitions between successive symbols as shown in Fig. 1.16. From (1.71) and Fig. 1.16 the instantaneous phase takes two values $(-\pi/2$ and $+\pi/2)$. For the 4-levels PSK, see Fig. 1.17.

3. Frequency shift keying (FSK)

The binary FSK signal is represented by [28]

$$s(t) = \cos(2\pi f_c t + D_f \int_{-\infty}^{t} m(\lambda)d\lambda) \tag{1.72}$$

where D_f is the modulation index of FSK2 signal. For binary FSK signal, it can be consider as two ASK2 signal centred at frequencies f_{mark} and f_{space}, defined later in chapter 3. Thus,

$$\begin{aligned} G_s(f) &= \frac{A^2}{16}[\delta(f - f_{mark}) + \delta(f + f_{mark})] \\ &+ \frac{A^2}{16}\left[\frac{\sin^2 \pi T_b(f - f_{mark})}{\pi^2 T_b(f - f_{mark})^2} + \frac{\sin^2 \pi T_b(f + f_{mark})}{\pi^2 T_b(f + f_{mark})^2}\right] \\ &+ \frac{A^2}{16}[\delta(f - f_{space}) + \delta(f + f_{space})] \\ &+ \frac{A^2}{16}\left[\frac{\sin^2 \pi T_b(f - f_{space})}{\pi^2 T_b(f - f_{space})^2} + \frac{\sin^2 \pi T_b(f + f_{space})}{\pi^2 T_b(f + f_{space})^2}\right], \tag{1.73} \end{aligned}$$

From (1.73) and Fig. 1.18, it is clear that the FSK2 signal spectrum appears like two ASK signals. Also, (1.72) can be re-expressed as

$$s(t) = Re\left\{\alpha(t)e^{j\omega_c t}\right\} \tag{1.74}$$

where

$$\alpha(t) = e^{j\phi(t)} \tag{1.75}$$

Thus, the instantaneous amplitude and phase are given by

$$a(t) = 1 \tag{1.76}$$

and

$$\phi(t) = D_f \int_{-\infty}^{t} m(\lambda)d\lambda \tag{1.77}$$

Although $m(t)$ is discontinuous at the bit transition, the instantaneous phase is continuous because ϕ (t) is proportional to the integral of $m(t)$. From (1.76), $a(t)$ is constant but due to the slope over detection phenomenon, a minor amplitude variation imposed as shown in Fig. 1.18. From (1.77) and Fig. 1.18, the instantaneous phase is time varying function. Furthermore, the instantaneous frequency is the bit stream (modulating signal). For 4-levels FSK, see Fig. 1.19.

1.4 Summary

In this chapter we have introduced some background to and motivations for modulation recognition research followed by some mathematical preliminaries needed for the book. Subsequently we have described different modulation types and their characteristics which will be useful in selecting features for automatic modulation recognition in Chapters 2 to 5.

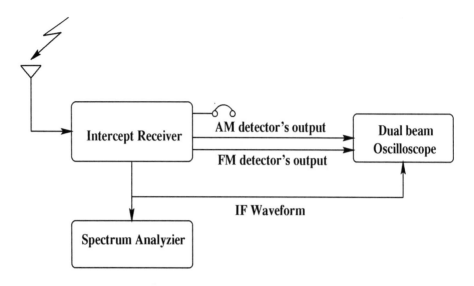

Figure 1.1: Equipment configuration for manual modulation recognition.

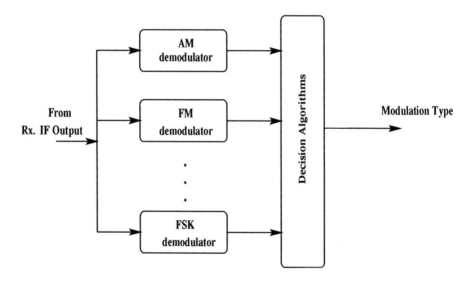

Figure 1.2: A modulation recogniser using a bank of demodulators and a set of decision algorithms [5].

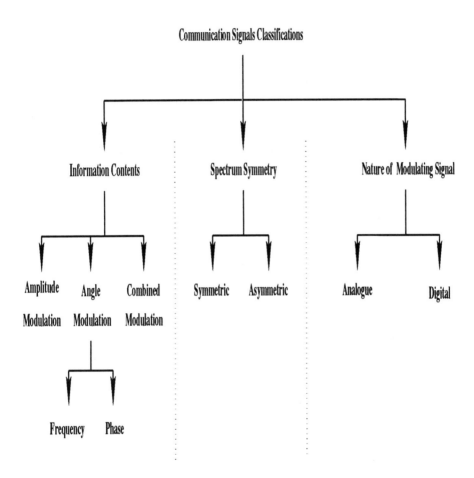

Figure 1.3: Communication signals classifications.

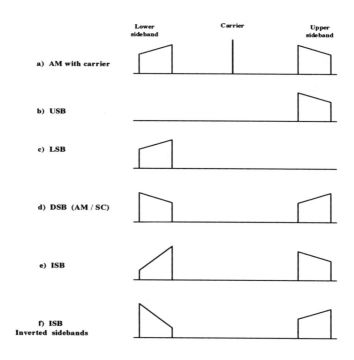

Figure 1.4: Spectrum shapes of amplitude modulated signal.

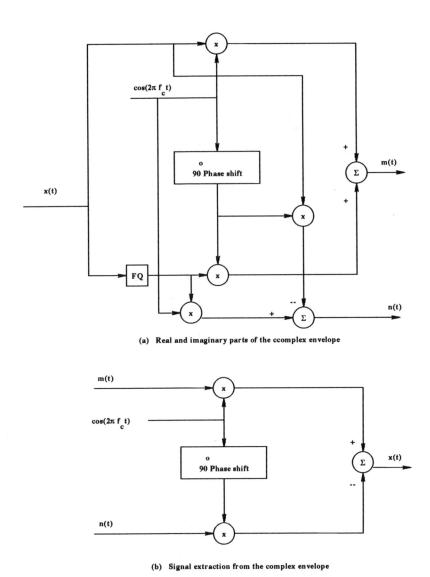

Figure 1.5: (a) Extraction of the real and imaginary parts of the complex envelope, (b) Extraction of the real signal from the real and imaginary parts of the complex envelope.

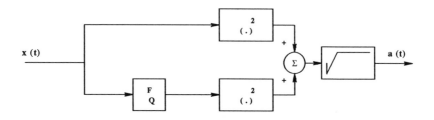

(a) **Extraction of the instantaneous amplitude**

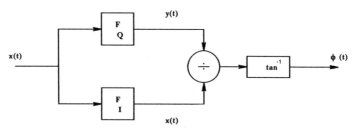

(b) **Extraction of the instantaneous phase**

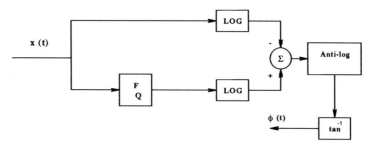

(c) **Practical implementation of the division operation**

Figure 1.6: Extraction of (a) the instantaneous amplitude and (b) the instantaneous phase, and (c) Practical implementation of the division operation.

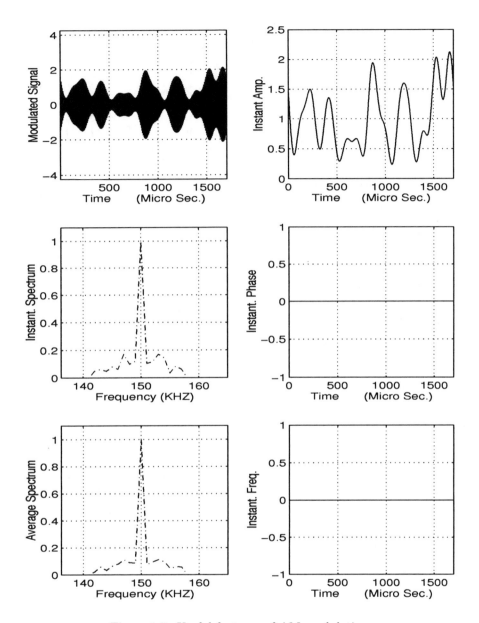

Figure 1.7: Useful features of AM modulation.

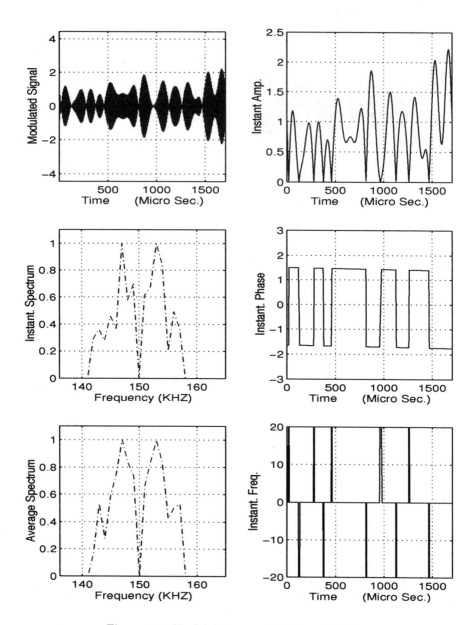

Figure 1.8: Useful features of DSB modulation.

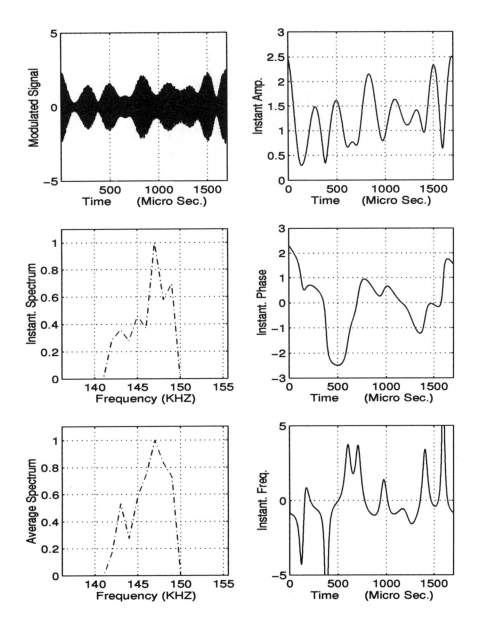

Figure 1.9: Useful features of LSB modulation.

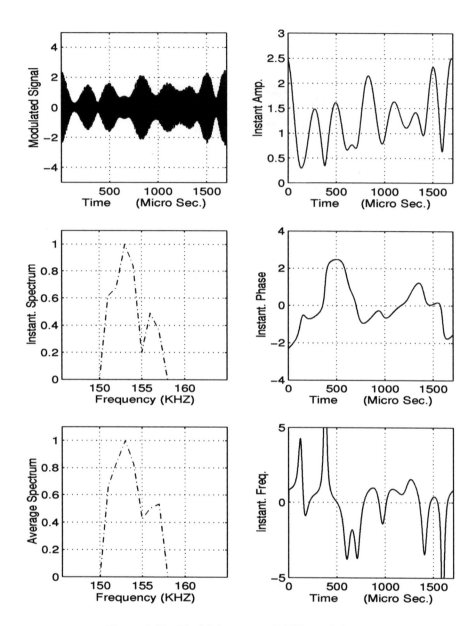

Figure 1.10: Useful features of USB modulation.

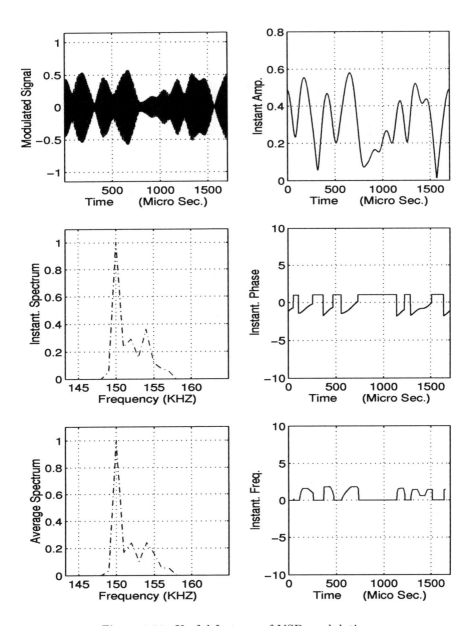

Figure 1.11: Useful features of VSB modulation.

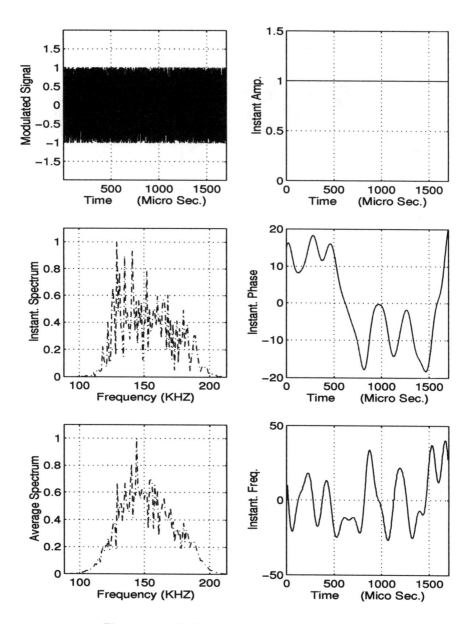

Figure 1.12: Useful features of FM modulation.

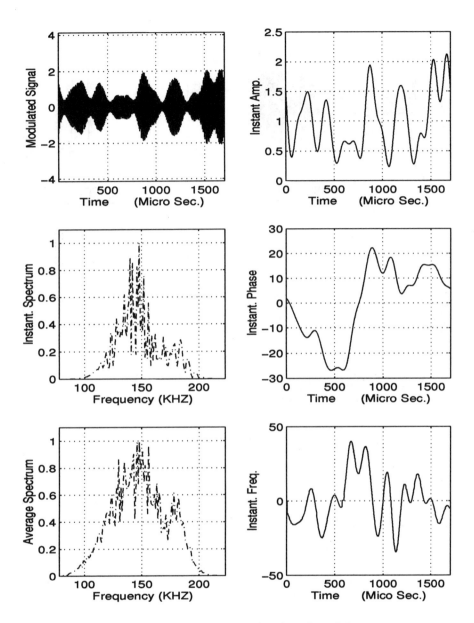

Figure 1.13: Useful features Combined modulation.

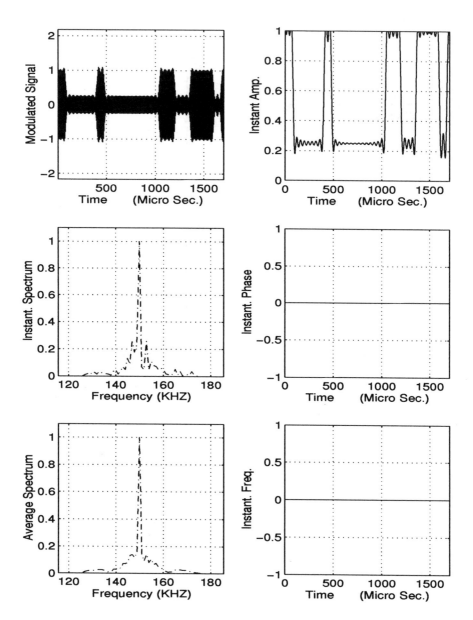

Figure 1.14: Useful features of ASK2 modulation.

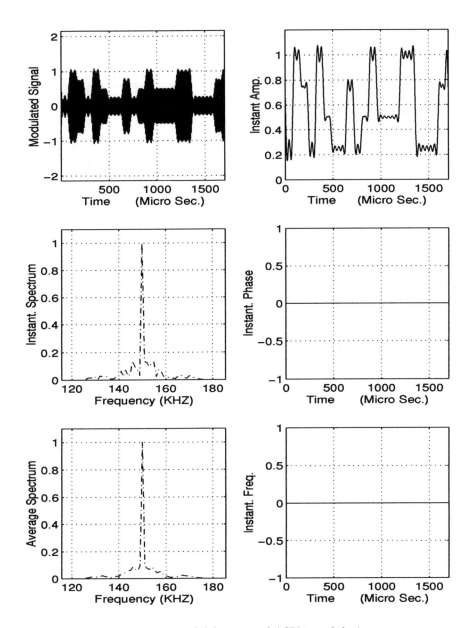

Figure 1.15: Useful features of ASK4 modulation.

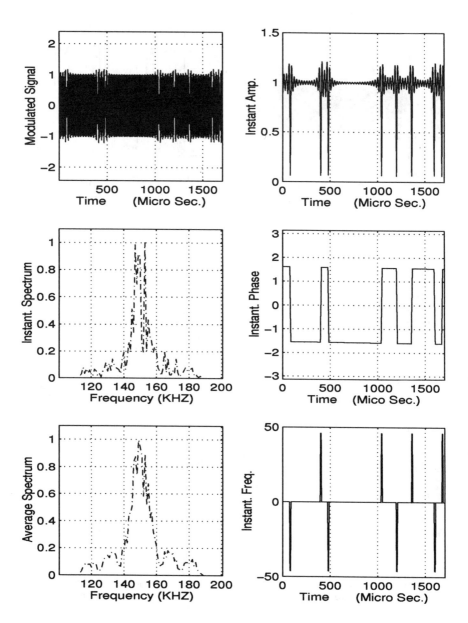

Figure 1.16: Useful features of PSK2 modulation.

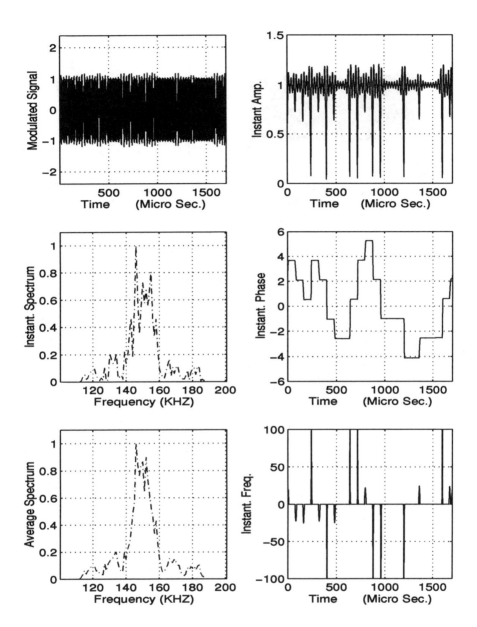

Figure 1.17: Useful features of PSK4 modulation.

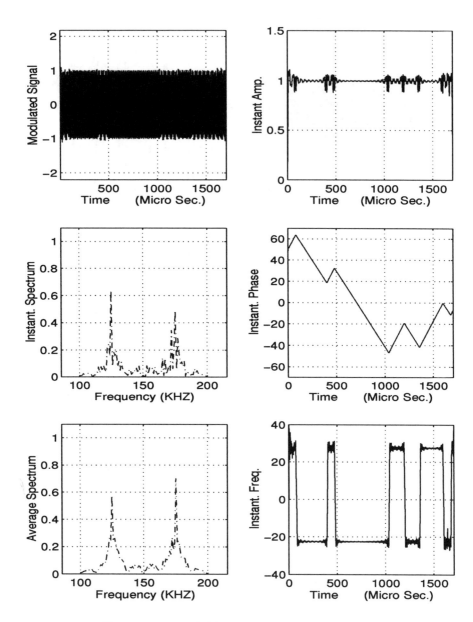

Figure 1.18: Useful features of FSK2 modulation.

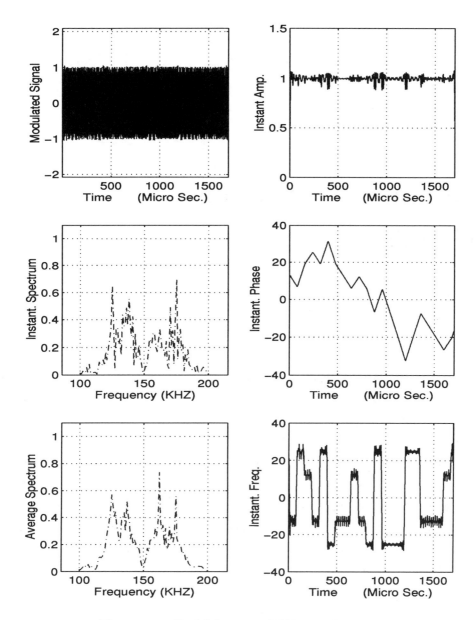

Figure 1.19: Useful features of FSK2 modulation.

Chapter 2

Recognition of Analogue Modulations

2.1 Introduction

The current trend in modern communication systems is the use of digital modulation types rather than the analogue ones. However, the analogue modulations are still in use especially in the third world countries. So, the aim of this chapter is to discuss some of the available analogue modulation recognisers and to introduce some quick and fast algorithms for the well known analogue modulation types. The analogue modulation types that can be classified with the developed algorithms, introduced in this chapter, are: AM (with different modulation depths (60% and 80%)), DSB, VSB, LSB, USB, FM (with different modulation indices (5 and 10)), and combined modulated signals (with different modulation depths and different modulation indices). Furthermore, a non-intelligible simulated speech signal is used as a modulating signal for simulating the different types of analogue modulations to increase the degree of realism.

Generally, any automatic modulation recogniser, based on the decision-theoretic approach, comprises three main stages: 1) pre-processing, 2) key feature extraction, and 3) modulation classification. From the modulation recognition point of view, the main functions of the pre-processing stage are the signal isolation (only one signal is considered for the modulation recognition process) and the signal segmentation. The segment length should be chosen to achieve two main requirements: 1) avoiding the fading modulation effects and 2) providing good features extraction. The proposed key features are extracted from each available segment of the intercepted signal frame. The extracted key features are compared with suitable thresholds according to set of decision rules to decide about the modulation type possessed in each segment, then a global decision is taken over all the available segments.

In this chapter a review of five analogue modulation recognisers is presented. A further five algorithms for analogue modulation recognition, based on the decision-theoretic approach, are introduced. These algorithms use the same key features that are extracted using the conventional signal processing tools. Thus, the developed algorithms can be implemented at extremely low cost as well as can be used in real field applications. In the next section, a review of the relevant previous work is introduced. In section 2.3, the proposed analogue modulation recognition algorithms along with the extraction of the required key features are introduced. The details of computer simulations for different types of band-limited analogue modulated signals of interest and those are used in the performance evaluations of the proposed algorithms are introduced in Section 2.4. The thresholds determination and the performance evaluations of the proposed algorithms are introduced in Section 2.5. Finally, the chapter is concluded in Section 2.6.

2.2   Relevant Previous Work

Fabrizi et al. [7] suggested a modulation recogniser for analogue modulations, based on the variations of both the instantaneous amplitude and the instantaneous frequency. The key features used are: 1) the ratio of the envelope peak to its mean, and 2) the mean of the absolute value of the instantaneous frequency. This recogniser is used to discriminate between some types of analogue modulation - CW, AM, FM, and SSB. This recogniser is tested by 24 realizations each with 250 msec. length for each modulation type of interest. In [7], it is claimed that with these two key features the discrimination of the AM and CW signals from the FM and SSB signals could be achieved at SNR $\geq$ 35 dB. However, the SSB could be separated from FM signals at SNR $\geq$ 5 dB.

Chan and Gadbois [10] proposed a modulation recogniser based on the envelope characteristics of the intercepted signal. This recogniser employs a technique in which the instantaneous amplitude of the intercepted signal is computed using a 31-coefficient Hilbert transformer. It uses the ratio R of the variance of the squared instantaneous amplitude to its mean square, as shown in Fig. 2.1, as a decision criterion to decide about the correct modulation type. This recogniser is used for the recognition of some analogue modulated signals - AM, FM, DSB and SSB. The choice for the ratio R to discriminate between these modulation types is based on the following fact: in noise-free signals, the ratio R is zero for FM signals and close to unity for AM signals. Furthermore, the ratio R for DSB signals is 2 and is 1 for SSB signals. The simulation results for the developed recogniser were derived from 200 realizations for each modulation type of interest, each with length 2048 samples (equivalent to 20 msec.) as is introduced in ([10], Table 3). In [10] it is claimed that at a SNR of 7 dB, the probability of correct modulation recognition is 100% for FM signals, 90.5% for AM signals, 80.0% for SSB signals and 94.0% for DSB signals. But *Azzouz* in [29] numerically analysed the mathematical expressions of the ratio R for different modulation types in [10] and it was found that this recogniser requires SNR $\geq$ 13.5 dB to distinguish between the amplitude signal and the non-amplitude signal with confidence level not less than 89.0%. Furthermore, in [10] nothing is mentioned about the recognition of the VSB and the combined modulated signals. From the author's point of view, this recogniser by its nature cannot recognise the signals that have both amplitude and phase information (combined modulated signals).

Jovanovic et al. [15] introduced a modulation recogniser to discriminate between a low modulation depth AM and a pure carrier wave (CW) in a noisy environment. The key feature used is defined as the ratio of the variance of the in-phase component to that of the quadrature component of the complex envelope of a signal. The only thing mentioned about the performance evaluation is that the proposed key feature is a highly reliable tool for separating the AM signals with low modulation depth from the unmodulated carrier even if the SNR is poor.

Nagy [19] proposed a modulation recogniser for analogue radio signals only. In this recogniser the *Chan and Gadbois* parameter [10], R, in addition to the variance of the instantaneous frequency normalised to the squared sample time, are used as key features to discriminate between the different modulation types of interest. The modulation types that can be classified by this recogniser are AM, DSB, SSB, FM, and CW. In [19], it is mentioned that the performance evaluation of this recogniser was derived from 500 realizations, each with 1024 samples (equivalent to 0.25 sec.), for each modulation type. Also, it is claimed that based on the ratio R [10] only, the different modulation types have been classified with success rate $\geq$ 90.0% at SNR = 15 dB except the SSB had been classified with success rate 66.0% as shown in [19]. So, *Nagy* suggested the second key feature, which is the variance of the instantaneous frequency normalised to the squared sampling time. In [19], it is claimed that using these two key features, the SSB has been classified with success rate $\geq$ 94.0% at SNR = 15 dB, and the other types have been classified with success rate 100%.

Al-Jalili [23] proposed a modulation recogniser to discriminate between the USB and LSB signals. This recogniser is based on the fact that the instantaneous frequency of the USB signal has more negative frequency spikes than positive ones, and vise versa the LSB signal. The key feature used in this recogniser is defined as the ratio, G, of the number of negative spikes to the number of the positive ones of the instantaneous frequency. So, G is > 1 for USB and it is < 1 for LSB. In [23] the performance measure of the proposed recogniser is derived from 10 realizations, each with 128 msec. for each modulation type and it is claimed that this recogniser performs well for SNR $\geq$ 0.

2.3 Developed Analogue Modulated Signal Recognition Algorithms (AMRAs)

In this section, a global procedure for band-limited analogue modulated signal recognition, based on the decision-theoretic approach, is proposed and analysed. Additionally, a criterion is developed to judge the quality of different segments of the intercepted signal frame according to their suitability for modulation recognition. First, the intercepted signal frame with length L seconds is divided into successive segments, each with length N_s = 2048 samples (equivalent to 1.707 msec.), resulting in $M_s = \frac{L f_s}{N_s}$ segments, where f_s (=1200 kHz) is the sampling rate. Thus, in the proposed algorithms for analogue modulation recognition, the decision about the modulation type is taken first from each available segment. Second, a global decision is determined from all the M_s available segments of the intercepted signal frame by comparing the global decision with each segment decision. So, the proposed algorithms for analogue modulation recognition require classification of each segment and classification of a signal frame. Similar ideas have been published in [30] and [31].

2.3.1 Classification of each segment

From every available segment, the suggested procedure to discriminate between the different types of analogue modulations comprises two steps: 1) key features extraction and 2) modulation classification.

Key feature extraction

In the proposed AMRAs, four key features are used to discriminate between the modulation types of interest and they are derived from the instantaneous amplitude $a(t)$ and the instantaneous phase $\phi(t)$ as well as the RF signal spectrum. For the mathematical expressions of the instantaneous amplitude and phase as well as the RF signal spectrum for different types of analogue modulations, see Section 1.3.1.

The first key feature, γ_{max}, is defined by

$$\gamma_{max} = \max \mid DFT\left(a_{cn}(i)\right) \mid^2 / N_s \tag{2.1}$$

where N_s is the number of samples per segment and $a_{cn}(i)$ is the value of the normalised-centred instantaneous amplitude at time instants $t = \frac{i}{f_s}, (i = 1, 2, ..., N_s)$, and it is defined by

$$a_{cn}(i) = a_n(i) - 1, \quad where \quad a_n(i) = \frac{a(i)}{m_a}, \tag{2.2}$$

where m_a is the average value of the instantaneous amplitude evaluated over one segment; i.e.

$$m_a = \frac{1}{N_s} \sum_{i=1}^{N_s} a(i), \tag{2.3}$$

Normalisation of the instantaneous amplitude is necessary to compensate the channel gain. Thus, γ_{max} represents the maximum value of the spectral power density of the normalised-centred instantaneous amplitude of the intercepted signal.

The second key feature, σ_{ap}, is defined by

$$\sigma_{ap} = \sqrt{\frac{1}{C}\left(\sum_{a_n(i)>a_t} \phi_{NL}^2(i)\right) - \left(\frac{1}{C}\sum_{a_n(i)>a_t} \mid \phi_{NL}(i) \mid\right)^2} \tag{2.4}$$

where $\phi_{NL}(i)$ is the value of the centred non-linear component of the instantaneous phase at time instants $t = \frac{i}{f_s}$, C is the number of samples in $\{\phi_{NL}(i)\}$ for which $a_n(i) > a_t$ and a_t is a threshold for $\{a(i)\}$ below which the estimation of the instantaneous phase is very sensitive to the noise. The determination of the this threshold will be explained later. Thus, σ_{ap} is the standard deviation of the absolute value of the centred non-linear component of the instantaneous phase, evaluated over the non-weak intervals of a signal segment.

The third key feature, σ_{dp}, is defined by

$$\sigma_{dp} = \sqrt{\frac{1}{C}\left(\sum_{a_n(i)>a_t}\phi_{NL}^2(i)\right) - \left(\frac{1}{C}\sum_{a_n(i)>a_t}\phi_{NL}(i)\right)^2} \tag{2.5}$$

Thus, σ_{dp} is the standard deviation of the centred non-linear component of the direct (not absolute) instantaneous phase, evaluated over the non-weak intervals of a signal segment.

The fourth key feature is used for measuring the spectrum symmetry around the carrier frequency, and it is based on the spectral powers for the lower and upper sidebands of the RF signal. So, it is defined as:

$$P = \frac{P_L - P_U}{P_L + P_U}, \tag{2.6}$$

where,

$$P_L = \sum_{i=1}^{f_{cn}} | X_c(i) |^2, \tag{2.7}$$

$$P_U = \sum_{i=1}^{f_{cn}} | X_c(i + f_{cn} + 1) |^2, \tag{2.8}$$

where, $X_c(i)$ is the Fourier transform of the RF signal, $x_c(i)$, $(f_{cn} + 1)$ is the sample number corresponding to the carrier frequency, f_c, and f_{cn} is defined as

$$f_{cn} = \frac{f_c N_s}{f_s} - 1 \tag{2.9}$$

A detailed pictorial representation for key feature extraction from a signal segment is shown in Fig. 2.2 in the form of a flowchart.

Modulation classification procedure

Based on the aforementioned four key features, many algorithms can be developed according to the time order of applying these key features in the classification procedure. In this chapter, five algorithms for analogue modulations recognition as shown in Figs. 2.3 and C.1 - C.4, using the same key features, are considered. The details of only one algorithm are presented in this section. Furthermore, the performance measures for all the analogue modulation types of interest are introduced at the SNR of 10 dB, 15 dB and 20 dB. Also, the overall success rates at the 10 dB, 15 dB, and 20 dB SNRs for the developed AMRAs are introduced in this chapter. Meanwhile, the other four algorithms along with their performance evaluations are introduced in Appendix C.1.

AMRA I

It is worth noting that each decision rule is applied to a set of modulation types separating it into two non over-lapping subsets - A and B. The choice of γ_{max}, σ_{ap}, σ_{dp} and the ratio P as key features for the proposed AMRA I is based on the following facts:

- γ_{max} is used to discriminate between FM signals as a subset and DSB, and combined (AM-FM) signals as the second subset. As the FM signals have constant instantaneous amplitude, their normalised-centred instantaneous amplitude is zero. Thus, their spectral power densities are also zero; i.e. they have no amplitude information ($\gamma_{max} < t(\gamma_{max})$). On the other hand, DSB, and combined (AM-FM) signals possess amplitude information ($\gamma_{max} \geq t(\gamma_{max})$). So this key feature can be used to discriminate between the signals that have amplitude information (DSB, and combined) and that do not have amplitude information (FM).

- σ_{ap} is used to discriminate between DSB signals as a subset and Combined (AM-FM) signals as the second subset. From (1.32) it is clear that the direct phase of a DSB signal - after removing the linear phase component due to the carrier frequency - takes on values of 0 and π, so its absolute value after centring is constant ($= \pi/2$) such that it has no absolute phase information ($\sigma_{ap} < t(\sigma_{ap})$). On the other hand the combined modulated signals have absolute and direct phase information ($\sigma_{ap} \geq t(\sigma_{ap})$). So, σ_{ap} can be used to discriminate between the types that have absolute phase information (combined) and that have no absolute phase information (DSB).

- σ_{dp} is used to discriminate between AM and VSB signals as a subset and DSB, LSB, USB, FM, and combined (AM-FM) signals as the second subset. AM and VSB have no direct phase information ($\sigma_{dp} < t(\sigma_{dp})$). On the other hand, the other types have direct phase information ($\sigma_{dp} \geq t(\sigma_{dp})$) by their natures. So, σ_{dp} can be used to discriminate between the types that have direct phase information (DSB, LSB, USB, FM, and Combined) and that have no direct phase information (AM and VSB).

- the ratio P is used to discriminate between the VSB and AM signals as well as to discriminate between the SSB (LSB and USB) as a subset and the DSB, FM and combined modulated signals as the second subset, since $\mid P \mid$ at infinite SNR ought to be **1** for SSB signals (+1 for LSB and -1 for USB), and **0** for AM, DSB, FM and combined modulated signals. It is well known that the VSB is an intermediate type between the AM and the SSB - USB - signals from the spectrum symmetry point of view. So, it is suggested that the threshold $t(P)$ is chosen between 0 and 1. Thus, by using this rule, it is possible to discriminate the VSB signal from the AM one as well as to discriminate the SSB from the DSB, FM, and combined (AM-FM) signals.

A detailed pictorial representation of the proposed analogue modulations recognition procedure is shown in Fig. 2.2 in the form of a flowchart.

2.3.2 Classification of a signal frame

As it is possible to obtain different classifications of the M_s segments (generated from a signal frame segmentation), the majority logic rule is applied; i.e. select the classification with largest number of repetitions. If two or more classifications have equal maximum numbers of repetitions, they are regarded as candidates for the optimal decision. In this case, continue as follows:

1. group the segments corresponding to each of the candidate decisions;

2. determine for every segment within a group the number of samples of the instantaneous amplitude falling below the threshold a_t. Evaluate the total numbers of these samples over the group; and

3. adopt the decision whose corresponding group has minimum number of samples falling below the threshold a_t.

It is worth noting that, due to the simplicity of both the key feature extraction as shown in Fig. 2.2 (using conventional signal processing tools only) and the decision rules as shown in Figs. 2.3 and C.1 - C.4 (simple logic functions), the proposed decision-theoretic AMRAs can be used for on-line analysis.

2.4 Computer Simulations

Because of the classified nature of the problem, the author found it difficult to obtain real modulated signals. So, in this section software generation of different types of analogue modulated signals with high degree of realism is introduced. Software generation of test signals provides a larger flexibility for selecting parameters and of adjusting specific values of the SNR. While the hardware generation of test signals does not provide the same degree of flexibility, it provides a larger degree of realism. Furthermore, in the primary phase of performance evaluation of any system or algorithm, computer simulations and software generation of test signals are essential.

Also, due to the excessively large number of computer simulations and because of the sharp time limitation for preparation of this book, it was necessary to assign some priorities to the required computer simulations. The main reasons for not deriving the performance data from a Monte-Carlo simulation trial are:

1. for each trial it is necessary to repeat the whole procedure described in Section 2.2 including the iterative determination of the thresholds, that will be explained later, which already consumes a long time, and

2. the decision at the end of the proposed algorithms is derived from several segments (classification of a signal frame) and not from a single segment.

This section is concerned with introducing complete software simulations and an extensive explanation for the generation of different types of band-limited analogue modulated signals corrupted with a band-limited Gaussian noise that are used in the performance evaluation.

2.4.1 Analogue modulated signal simulations

The carrier frequency, f_c, and the sampling rate, f_s, were respectively chosen equal to 150 kHz and 1200 kHz. In order to increase the degree of realism, a non-intelligible simulated speech signal is used as a modulating signal for all analogue modulation types of interest. The procedure used for generation of a non-intelligible speech signal of length 1.707 msec. (corresponding to 2048 samples) comprises the following steps:

1. generation of a sequence $\{n(i)\}$ of N_s independent values from a zero-mean uniformly distributed random number generator.

2. calculation of another sequence $\{x_m(i)\}$ according to the difference equation representing a first order auto-regressive model given by

$$x_m(i) = \rho x_m(i-1) + n(i), \qquad x_m(0) = 0 \qquad (2.10)$$

where the coefficient ρ was chosen such that the -3 dB spectrum bandwidth of the simulated auto-regressive signal was equal to 4 kHz.

3. evaluation of the N_s-point FFT of the sequence $\{x_m(i)\}$ to be able to simulate a low pass filter with cut-off frequency equal to 8 kHz.

4. evaluation of the N_s-point inverse FFT.

The normalised spectral power density of the first order auto-regressive model (2.10) is given by [32]

$$\gamma\left(e^{\frac{j2\pi f}{f_s}}\right) = \frac{(1-\rho)^2}{1+\rho^2 - 2\rho\cos(\frac{2\pi f}{f_s})} \qquad (2.11)$$

Thus, by substituting $f = 4$ kHz and $f_s = 1200$ kHz in (2.11) and equating the right hand side to $\frac{1}{2}$, it was found that ρ ought to be 0.98.

AM, DSB, FM and combined (AM-FM) signals were derived from the general expression [27],

$$s(t) = (A + m x_1(t)) \cos\left[2\pi f_c t + K_f \int_{-\infty}^{t} x_2(\tau) d\tau\right], \qquad (2.12)$$

where $x_1(t)$ and $x_2(t)$ are two modulating signals (simulated speech signals), m is a coefficient determined by the desired amplitude modulation depth (2.13), and K_f is a frequency deviation coefficient (2.14). Thus

$$Q = \frac{m \{[x_1(t)]_{max} - [x_1(t)]_{min}\}}{2A + m \{[x_1(t)]_{max} + [x_1(t)]_{min}\}}, \qquad (2.13)$$

where Q is the desired amplitude modulation depth, and

$$K_f = \frac{2\pi f_x D}{[x_2(t)]_{max}}, \qquad (2.14)$$

where D is the desired frequency modulation index, and f_x is the maximum frequency in the spectrum of the modulating signal ($= 8$ kHz). The integral in (2.12) was numerically evaluated using the trapezoidal rule as follows:

$$x_{int}(i) = \frac{1}{2f_s} \left[x_2(1) + x_2(i) + 2 \sum_{j=2}^{i-1} x_2(j) \right]. \qquad (2.15)$$

Thus, by a suitable selection of the parameters A, m, and K_f in (2.12), any one of the aforementioned analogue modulation types can be generated. The corresponding values of these parameters for each of these modulation types are shown in Table 2.1.

SSB signals were generated according to the expression [27]

$$s(t) = x(t) \cos(2\pi f_c t) \pm y(t) \sin(2\pi f_c t) \qquad (2.16)$$

where $x(t)$ is a simulated speech signal and $y(t)$ is its Hilbert transform. The minus sign in (2.16) is used for USB signal simulation whereas the positive sign is used for LSB signal simulation [27].

VSB signals were derived from AM signals using a VSB filter [27]. The analytic expression of the magnitude and phase responses of the VSB filter used in our simulations are given by

$$| H_{VSB}(f) | = \begin{cases} (1/2\alpha) [f - (f_c - \alpha)] & \text{if } f_c - \alpha \leq f < f_c + \alpha \\ 1 & \text{if } f_c + \alpha \leq f \leq f_c + f_x \\ 0 & \text{otherwise} \end{cases} \qquad (2.17)$$

$$\phi_{VSB}(f) = \begin{cases} (-\pi/\alpha) [f - (f_c - \alpha)] + \pi & \text{if } f_c - \alpha \leq f < f_c + \alpha \\ [-2\pi/(f_x - \alpha)] [f - (f_c - \alpha)] + \pi & \text{if } f_c + \alpha \leq f \leq f_c + f_x \\ 0 & \text{otherwise} \end{cases} \qquad (2.18)$$

where f_x is the maximum frequency of the simulated speech signal, and α is chosen such that $\frac{2\alpha}{f_c} \geq 0.01$ [27]. In our simulations $\alpha = 2$ kHz at $f_c = 150$ kHz.

2.4.2 Band-limiting of simulated modulated signals

Every communication transmitter has a finite transmission bandwidth. Consequently, the transmitted signal is band-limited. Therefore, the simulated modulated signals were band-limited in order to make them represent more realistic test signals for the proposed global procedure for analogue modulation recognition. The band-limitation of the simulated analogue modulated signals was carried out in accordance with the usual implementation in practice. Thus, in analogue modulated signals the band-limitation was exercised on the modulating signal (non-intelligible speech signal) as well as on the modulated signals. Finally, the bandwidths of the simulated analogue modulated signals are presented in Table 2.2. Typical examples of the modulating signals, modulated signals and the values of all the aforementioned four key features for each analogue modulation type of interest based on 1.707 msec. duration signal (equivalent to 2048 samples) are shown in Figs. 2.4 - 2.10.

2.4.3 Noise simulation and SNR adjustment

The noise sequence used in the performance evaluation of the proposed AMRAs is a band-pass Gaussian noise. The procedure for generating this noise sequence are as follows:

1. generation of a Gaussian noise sequence $\{n(i)\}$ with length equal to 2048 independent values of a random number generator with zero-mean, and normally distributed; and

2. band-pass filtering of the sequence $\{n(i)\}$ using a band-pass filter with bandwidth related to the intended type of modulation to generate another sequence $\{n_n(i)\}$, which is used in performance evaluations.

Usually in practice, the bandwidth of any intercepting receiver is chosen to be slightly larger than the intercepted signal bandwidth. So, in order to increase the degree of realism of the simulated additive Gaussian noise, it was band-limited to a bandwidth equal to 1.2 times the simulated modulated signal bandwidth.

Finally, any desired SNR was adjusted by multiplying the generated band-limited Gaussian noise sequence $\{n_n(i)\}$ by a coefficient R_{sn} determined from

$$R_{sn} = \sqrt{\frac{S_p}{N_p}} \left[10^{\frac{-SNR}{20}} \right] \tag{2.19}$$

where SNR is substituted in decibels, and

$$S_p = \sum_{i=1}^{N} s^2(i), \tag{2.20}$$

and

$$N_p = \sum_{i=1}^{N} n_n^2(i) \tag{2.21}$$

2.5 Thresholds Determinations and Performance Evaluations

2.5.1 Determination of the relevant thresholds

The implementation of the proposed AMRAs, introduced in section 2.2 and Appendix C.1, requires the determination of four key feature thresholds: $t(\gamma_{\max})$, $t(\sigma_{ap})$, $t(\sigma_{dp})$, and $t(P)$ in addition to the normalised amplitude threshold, $a_{t_{opt}}$, which is used in measuring the key features σ_{ap} and σ_{dp} as well as in the classification of a signal frame. As a result of the signal segmentation in the pre-processing stage (the available signal frame is divided into M_s successive segments), the threshold determination and the performance evaluations are derived from 400 realizations, each with 2048 samples (equivalent to 1.707 msec.), for each modulated signal of interest at the SNR of 10 dB and 20 dB.

Determination of the optimum key features threshold values

The determination of the optimum key features threshold values is derived from 400 realizations for each modulated signal of interest at the SNR of 10 dB and 20 dB. A global procedure for key feature threshold determination is proposed and analysed. From Figs. 2.3 and C.1 - C.4, it is clear that each decision rule is applied to a set of modulation types, G, separating it into two non-overlapping subsets (A and B) according to

$$KF \mathrel{\mathop{\gtrless}^{A}_{B}} x_{opt} \tag{2.22}$$

where KF is the measured value of the chosen key feature and x_{opt} is the corresponding optimum threshold value. In Fig. 2.11, each sub-plot contains four curves: 1) the dotted curve represents the conditional probability $P(B(x)/B)$, 2) the dashdotted curve represents the conditional probability $P(A(x)/A)$, 3) the dashed curve is used to determine the optimum key feature threshold, x_{opt}, as

$$x_{opt} = arg \min_x \{K(x)\} \tag{2.23}$$

where

$$K(x) = \frac{P(A(x)/B)}{P(A(x)/A)} + \frac{P(B(x)/A)}{P(B(x)/B)} + \mid 1 - P(A(x)/B) - P(A(x)/A) \mid \tag{2.24}$$

and 4) the solid curve is used to determine the average probability of correct decisions about the two subsets, A and B, at the optimum threshold, x_{opt}, as

$$P_{av}(x_{opt}) = \frac{P(A(x_{opt})/A) + P(B(x_{opt})/B)}{2} \tag{2.25}$$

The optimum values for the key features thresholds, $t(\gamma_{max})$, $t(\sigma_{ap})$, $t(\sigma_{dp})$ and $t(P)$, and the corresponding average probability of correct decisions, $P_{av}(x_{opt})$, (based on the 400 realizations at the SNR of 10 dB and 20 dB and for the twelve analogue modulated signals) are shown in Table 2.3 and they can be extracted from Fig. 2.10. It is clear from Table 2.3 and Fig. 2.11 that the optimum thresholds values for some key features sometimes take regions instead a single value. In the performance evaluation only one definite value for each key feature threshold is used.

The optimum values of $t(\gamma_{max})$, $t(\sigma_{ap})$, $t(\sigma_{dp})$, and $t(P)$, are chosen to be 6, $\pi/4$, $\pi/6$, and 0.6 for VSB and 0.5 for SSB respectively. These optimum values are chosen to measure the performance of the analogue modulation recognition algorithm I. Furthermore, the results of the dependence of the aforementioned key features on SNR for only one realization of different modulation types of interest, for the AMRA I are shown in Figs. 2.12 - 2.16. Sample results corresponding to the AMRA I at the SNR of 10 dB, 15 dB and 20 dB are presented in this section. Moreover, the overall success rates for the five algorithms at the SNR of 10 dB, 15 dB, and 20 dB are introduced. Meanwhile, the performance evaluations of the other four algorithms are introduced in Appendix C.1.

A- Dependence of $\gamma_{\max}$ on SNR

The dependence of γ_{max} on the SNR was computed for seven analogue modulated signals - DSB, combined (AM-FM) (Q=60% and D=5), combined (AM-FM) (Q=60% and D=10), combined (AM-FM) (Q=80% and D=5), combined (AM-FM) (Q=80% and D=10), FM (D=5) and FM (D=10). Of course, only two of these seven signals have no amplitude information and they are FM (D=5) and FM (D=10) signals. Thus, these seven types can be divided into two subsets, A and B. The subset A contains all the types that have amplitude information ($\gamma_{max} \geq t(\gamma_{max})$). These types are: DSB, combined (AM-FM) (Q=60% and D=5), combined (AM-FM) (Q=60% and D=10), combined (AM-FM) (Q=80% and D=5), combined (AM-FM) (Q=80% and D=10). The subset B contains all the types that have no amplitude information ($\gamma_{max} < t(\gamma_{max})$). These types are: FM (D=5) and FM (D=10).

The results for one simulation of these seven analogue modulated signals are presented in Fig. 2.12. From Fig. 2.12, it is observed that for SNR ≥ 7 dB the curve corresponding to FM (D=5) signal falls below the threshold level $t(\gamma_{max}) = 6$, and the curve corresponding to FM (D=10) signal falls below the threshold level $t(\gamma_{max}) = 6$ for SNR ≥ 0 dB. Furthermore, for SNR ≥ 0 dB the curves corresponding to the other five modulated signals are above the threshold $t(\gamma_{max}) = 6$.

B- Dependence of σ_{ap} on SNR

The dependence of σ_{ap} on the SNR was computed for five analogue modulated signals - DSB, combined (AM-FM) (Q=60% and D=5), combined (AM-FM) (Q=60% and D=10), combined (AM-FM) (Q=80% and D=5), combined (AM-FM) (Q=80% and D=10). Only one of these five signals have no absolute phase information and this is the DSB signal. Thus, these five types of modulation can be divided into two subsets, A and B. The subset A contains all the types that have absolute phase information ($\sigma_{ap} \geq t(\sigma_{ap})$). These types are: combined (AM-FM) (Q=60% and D=5), combined (AM-FM) (Q=80% and D=5), combined (AM-FM) (Q=60% and D=10), combined (AM-FM) (Q=80% and D=10). The subset B contains the DSB modulation type that has no absolute phase information ($\sigma_{ap} < t(\sigma_{ap})$).

The results for one simulation of these five analogue modulated signals are presented in Fig. 2.13. From Fig. 2.13, it is clear that for SNR $\geq$ 0 dB the curve corresponding to the DSB falls below the threshold level $t(\sigma_{ap}) = \pi/4$. Furthermore, for SNR $\geq$ 0 dB the curves corresponding to the other four modulated signals are above the threshold $t(\sigma_{ap}) = \pi/4$.

C- Dependence of σ_{dp} on SNR

The dependence of σ_{dp} on the SNR was computed for twelve analogue modulated signals - AM (Q=60%), AM (Q=80%), DSB, VSB, LSB, USB, FM(D=5), FM (D=10), combined (AM-FM) (Q=60% and D=5), combined (AM-FM) (Q=60% and D=10), combined (AM-FM) (Q=80% and D=5) and combined (AM-FM) (Q=80% and D=10). Of course three of these twelve signals have no direct phase information and they are AM (Q=60%), AM (Q=80%) and VSB signals. Thus, these twelve modulation types can be divided into two subsets, A and B. The subset A contains all the types that have direct phase information ($\sigma_{dp} \geq t(\sigma_{dp})$). These types are: DSB, LSB, USB, FM (D=5), FM (D=10), combined (AM-FM) (Q=60% and D=5), combined (AM-FM) (Q=60% and D=10), combined (AM-FM) (Q=80% and D=5) and combined (AM-FM) (Q=80% and D=10) signals. The subset B contains all the types that have no direct phase information ($\sigma_{dp} < t(\sigma_{dp})$). These types are: AM (Q=60%), AM (Q=80%) and VSB.

The results for one simulation of these twelve analogue modulated signals are presented in Fig. 2.14. From Fig. 2.14, it is clear that for SNR $\geq$ 2 dB the curves corresponding to AM (Q=60%) and AM (Q=80%) signals fall below the threshold level $t(\sigma_{dp}) = \pi/6$. Furthermore, for SNR $\geq$ 5 dB the curve corresponding to the VSB signal falls below the threshold $t(\sigma_{dp}) = \pi/6$. On the other hand, for SNR $\geq$ 0 dB the curves corresponding to the other signals are above the threshold $t(\sigma_{dp}) = \pi/6$.

D- Dependence of the ratio P on SNR

It is well known that the absolute value of the ratio P, defined by (2.6), ought to be zero for AM and DSB modulated signals (perfectly symmetric spectrum signals), and one for SSB (LSB and USB) modulated signals. Furthermore, VSB is an intermediate type between the AM and the SSB from the spectrum symmetry point of view.

Thus, for VSB modulated signal $0 <| P |< 1$. In the proposed algorithm for analogue modulation recognition introduced in section 2.2 (AMRA I), the absolute value of the ratio P is used in two places: 1) the discrimination of VSB from AM signals, and 2) the discrimination of SSB (LSB and USB) from DSB, FM, and combined modulated signals.

For the discrimination of the VSB from the AM types, the dependence of the ratio P on the SNR was computed for three analogue modulated signals - AM (Q=60%), AM (Q=80%) and VSB. Of course only the VSB modulated signal has spectrum asymmetry and the other types have spectrum symmetry. Thus, these three types can be divided into two subsets, A and B. The subset A contains the VSB modulation type that has spectrum asymmetry $(| P |\geq t(P))$. The subset B contains the other types that have spectrum symmetry $(| P |< t(P))$. These types are: AM (Q=60%) and AM (Q=80%) signals.

The results for one simulation of these three analogue modulated signals are presented in Fig. 2.15. From Fig. 2.15, it is clear that for SNR ≥ 0 dB the curves corresponding to AM (Q=60%) and AM (Q=80%) signals represent that $| P |$ is below the threshold level $t(P) = 0.6$. For SNR ≥ 0 dB the curve corresponding to the VSB signal represents that $| P |$ is above the threshold $t(P) = 0.6$.

For the discrimination of SSB from DSB, FM and combined modulated signals, the dependence of the ratio P on the SNR was computed for nine analogue modulated signals - LSB, USB, DSB, FM (D=5), FM (D=10), combined (AM-FM) (Q=60%, D=5), combined (AM-FM) (Q=60%, D=10), combined (AM-FM) (Q=80%, D=5) and combined (AM-FM) (Q=80%, D=10). Of course only the LSB and the USB modulated signals have spectrum asymmetry and the other modulated signal types have spectrum symmetry. Thus, these nine modulated signals can be divided into two subsets A and B. The subset A contains all the types that have spectrum asymmetry $(| P |\geq t(P))$. These types are: LSB and USB signals. The subset B contains all the types that have spectrum symmetry $(| P |< t(P))$. These types are: DSB, FM (D=5), FM(D=10), combined (AM-FM) (Q=60%, D=5), combined (AM-FM) (Q=60%, D=10), combined (AM-FM) (Q=80%, D=5) and combined (AM-FM) (Q=80%, D=10) signals.

The results for one simulation of these nine analogue modulated signals are presented in Fig. 2.16. From Fig. 2.16, it is clear that for SNR > 0 dB the curves corresponding to the FM, and the combined modulated signals represents that $\mid P \mid$ is below the threshold level $t(P) = 0.5$. Furthermore, for SNR > 0 dB the curves corresponding to the LSB and the USB signals represent that $\mid P \mid$ is above the threshold $t(P) = 0.5$.

Determination of the normalised amplitude threshold $a_{t_{opt}}$

The normalised amplitude threshold, $a_{t_{opt}}$, (not to be confused with the threshold of amplitude information key feature $t(\gamma_{\max})$) was determined according to the following steps [33]:

1. A range of normalised-instantaneous amplitudes comprising the optimum threshold $a_{t_{opt}}$ was first determined. As we know, there are several modulation types (AM with high modulation depth and DSB) affected mainly by the noise in the weak intervals of a signal; i.e. this phenomenon occurs at the phase reversals of a signal. Thus, it is not advantageous to choose the lower limit of the instantaneous amplitude to be zero. The lower bound, a_l, of this range of the normalised instantaneous amplitudes was assigned the value 0.4 which was chosen by intuition. It was felt that any further decrease of this value would allow more noisy samples to contribute to the value of σ_{ap}. Anyway, the obtained optimum threshold $a_{t_{opt}}$ validated the author's intuition i.e. $a_{t_{opt}}$ was found to be larger than 0.4. The upper bound, a_u, of the selected range of the normalised instantaneous amplitudes was determined by first choosing an upper bound very close to the selected lower bound. Then this upper bound was gradually raised by small increments and the correctness of the decision about the existence of phase information for the no-amplitude information signals (FM, D=5 and FM, D=10) was simultaneously checked with respect to noise-free signals till the first wrong decision (no absolute phase information).

2. For simplicity the simulated AM(Q=60%), AM(Q=80%), VSB, and DSB signals were respectively designated by the values 1,2,3, and 4 of the subscript i. Clearly, all these signals have no absolute phase information. Let $\rho_i(a_t)$ denote the threshold SNR imposed by the decision rule about σ_{ap} for these types with respect to the ith modulated signal when the normalised amplitude threshold is a_t. Above this SNR threshold, there is a correct decision about σ_{ap} for AM, DSB, and VSB

signals - correct decision means no absolute phase information ($\sigma_{ap} < t(\sigma_{ap})$). Below the SNR threshold it considers the phase variation, originated by the additive noise, as if they were due to signal and thus provides false decisions. The optimal threshold $a_{t_{opt}}$ is determined according to a minimax strategy.

$$F(a_t) = \max_{i_u \geq i \geq i_l} \rho_i(a_t), \quad a_u \geq a_t \geq a_l \tag{2.26}$$

$$a_{t_{opt}} = arg \min_{i_u \geq i \geq i_l} F(a_t) \tag{2.27}$$

where for the considered analogue modulations $i_l = 1$ and $i_u = 4$. In order to speed up the implementation of this procedure, the bisection method [34] was used to determine $\rho_i(a_t)$ for every pair (i, a_t). Also the golden section method [34] was used to rapidly locate the minimum of $F(a_t)$.

Thus, the determination of the optimum normalised amplitude threshold is determined by repeating the aforementioned procedure and choosing the threshold values that achieve maximum probability of correct modulation recognition. It was found that the optimum normalised amplitude threshold range is $a_{t_{opt}} = [0.858:1]$. In the proposed AMRAs $a_{t_{opt}}$ is set to be **1**.

2.5.2 Performance evaluations

The results of the performance evaluations of the proposed procedure for the AMRA I are derived from 400 realizations for each modulated signal using the pre-chosen optimum key features threshold values. The optimum key features threshold values, $t(\gamma_{max})$, $t(\sigma_{ap})$, $t(\sigma_{dp})$, and $t(P)$ are chosen to be 6, $\pi/4$, $\pi/6$, and 0.6 for VSB and 0.5 for SSB. Twelve analogue modulated signals have been simulated and sample results at three different SNRs corresponding to 10 dB, 15 dB and 20 dB are presented in Tables 2.4, 2.5 and 2.6 respectively. Each of the twelve analogue modulated signals at each SNR was simulated 400 times. Consider Table 2.4 for example, it can be observed that seven analogue modulation types (12 modulated signals) have been correctly classified with more than 97.0% success rate. Indeed four of these seven modulation types (nine modulated signals) have been successfully classified every time (100% success rate). The results in Table 2.6 corresponds to the SNR of 20 dB. It is clear that the success rate for correct modulation recognition has increased with increasing SNR, and now all the modulation types have been classified with success rate > 98.0%. Furthermore, five of the seven modulation types (ten of the twelve analogue modulated signals) have been classified with success rate 100%.

2.5.3   Processing Time and Computational Complexity

As it is important to know if the AMRAs - developed by the authors - will be suitable for real time analysis or not, the processing time and computational power were measured using the MATLAB software on the SUN SPARC station **20**. The values of the processing time (measured in seconds) and the computational power (measured in Megaflops) measured on the SUN SPARC station **20** are shown in Table 2.7. The processing time and the number of Megaflops, required to take a decision about the modulation type, correspond to only one signal segment in each case. The numbers in Table 2.7 are the average values of the measurements for 10 different realizations of each modulation type of interest at ∞ SNR.

2.6   Conclusions

The aim of the AMRAs - developed by the authors - has been to recognise automatically the types of analogue modulations in communication signals. The current approach has been to carry out this task using the decision-theoretic approach. A number of new key features is proposed to fulfil the requirement of these algorithms. Moreover, there are three motivations for using the proposed algorithms in the on-line analysis. These are: 1) simplicity of the key feature extraction, and as is clear, all the key features used are extracted using the conventional signal processing tools; 2) determination of the optimum threshold value for each key feature should be finished beforehand; i.e. finish the determination of the key features thresholds before using the proposed algorithms in the on-line analysis; and 3) the simplicity of the decision rules used in the decision about the modulation type and it is clear that each decision rule used in these algorithms comprises only one logic (comparison) function (**IF ... THEN ... ELSE IF ... END**) and none of them comprises a combination of more than one function. Finally, it is worth noting that similar ideas have been published in [26].

Extensive simulations of twelve analogue modulated signals have been carried out at different SNR. A procedure for threshold determination is introduced. The threshold determination and the performance evaluations are derived from 400 realizations for each modulated signal of interest at each SNR value (10 dB and 20 dB). Sample results have been presented at the SNR of 10 dB, 15 dB and 20 dB. It is found that the threshold SNR for correct analogue modulation recognition is about 10 dB, which is an improvement in the SNR threshold over previous results reported in [7], [10] and [19]. Also presented are measured processing times and required computational power of the proposed AMRA I, that used for the discrimination of the different types of analogue modulations considered in the book.

In the AMRAs -developed by the authors - the key features threshold values are derived from 400 realizations of each modulated signal of interest at the SNR of 10 dB and 20 dB. The overall success rates for the five algorithms at three different SNRs (10 dB, 15 dB and 20 dB) and derived from 400 realizations for each of the twelve analogue modulated signals are shown in Table 2.8. It is found that the overall success rate of AMRA V is exactly the same as AMRA I for all the SNR values (10dB, 15 dB, and 20 dB). Also, the overall success rates for AMRA II and AMRA III are exactly the same. Generally, it is concluded that based on the same key features, different algorithms can be generated according to the sequence of applying the proposed key features in the classification algorithm and they perform with different success rates at the same SNR. In the decision-theoretic algorithms, the probability of correct decision about a modulation type is based on the the time-order of applying the set of decision rules in the classification algorithm as well as the probability of correct decision derived from each key feature. This implies the need of a new method in which the discrimination between the different modulation types should be derived from all the key features simultaneously; i.e. removing the effect of the time-ordering.

Modulation Type	A	m	K_f
AM	1	$\neq 0$ (evaluated from (2.13))	0
DSB	0	1	0
FM	1	0	$\neq 0$ (defined by (2.14))
Combined	1	$\neq 0$ (evaluated from (2.13))	$\neq 0$ (defined by (2.14))

Table 2.1: Analogue modulated signals parameter selection.

Modulation Type	Theoretical expression	Simulated values (kHz)
AM, DSB	$2 f_x$	16
VSB	$f_x + \alpha$	10
SSB	f_x	8
Combined	$2(D + 2) f_x$	112 if D = 5
		192 if D = 10
FM	$2(D + 1) f_x$	96 if D = 5
		176 if D = 10

Table 2.2: Bandwidths of analogue modulated signals.

Key Feature Thresholds	Optimum Value	$P_{av}(x_{opt})$	Notes
$t(\gamma_{max})$	[5.5-6]	100%	
$t(\sigma_{ap})$	$[\pi/6.5 - \pi/2.5]$	100%	
$t(\sigma_{dp})$	$\pi/6$	99.5%	
$t(P)$	[0.5-0.99]	100%	SSB
	[0.55-0.6]	100%	VSB

Table 2.3: Optimum key feature threshold values for the AMRA I.

Simulated	Deduced Modulation Type						
Modulation Type	AM	DSB	VSB	LSB	USB	COM.	FM
AM	100%	-	-	-	-	-	-
DSB	-	100%	-	-	-	-	-
VSB	-	-	98.0%	-	2.0%	-	-
LSB	-	-	0.2%	99.8%	-	-	-
USB	-	-	2.2%	-	97.8%	-	-
COM.	-	-	-	-	-	100%	-
FM	-	-	-	-	-	-	100%

Table 2.4: Confusion matrix for the AMRA I [based on 400 realizations] at SNR = 10 dB.

Simulated	Deduced Modulation Type						
Modulation Type	AM	DSB	VSB	LSB	USB	COM.	FM
AM	100%	-	-	-	-	-	-
DSB	-	100%	-	-	-	-	-
VSB	-	-	100%	-	-	-	-
LSB	-	-	-	100%	-	-	-
USB	-	-	1.7%	-	98.3%	-	-
COM.	-	-	-	-	-	100%	-
FM	-	-	-	-	-	-	100%

Table 2.5: Confusion matrix for the AMRA I [based on 400 realizations] at SNR = 15 dB.

Simulated	Deduced Modulation Type						
Modulation Type	AM	DSB	VSB	LSB	USB	COM.	FM
AM	100%	-	-	-	-	-	-
DSB	-	100%	-	-	-	-	-
VSB	-	-	100%	-	-	-	-
LSB	-	-	0.5%	99.5%	-	-	-
USB	-	-	1.2%	-	98.8%	-	-
COM.	-	-	-	-	-	100%	-
FM	-	-	-	-	-	-	100%

Table 2.6: Confusion matrix for the AMRA I [based on 400 realizations] at SNR = 20 dB.

Modulation Type	Computational power (Megaflops)	Processing time (Sec.)
AM	0.42	5.90
DSB	0.50	5.87
VSB	0.45	5.98
USB	0.51	5.93
LSB	0.51	5.94
Combined	0.51	5.97
FM	0.43	6.02

Table 2.7: Measured computational power and processing times of the AMRA I on SPARC 20.

SNR (dB)	Overall success rates		
	I & V	II & III	IV
10	99.4%	97.2%	98.0%
15	99.8%	97.5%	98.5%
20	99.8%	97.7%	98.8%

Table 2.8: Overall success rates for the developed AMRAs.

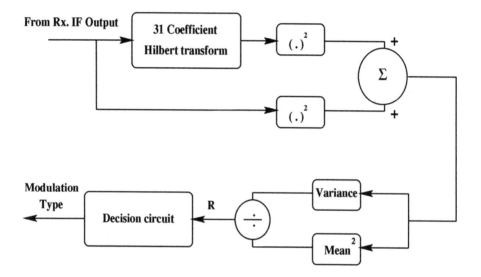

Figure 2.1: Simplified block scheme of the recogniser using the envelope characteristic only [10].

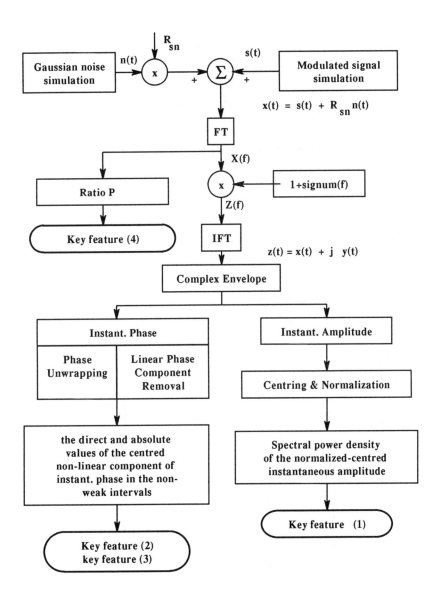

Figure 2.2: Functional flowchart for key features extraction in the AMRAs.

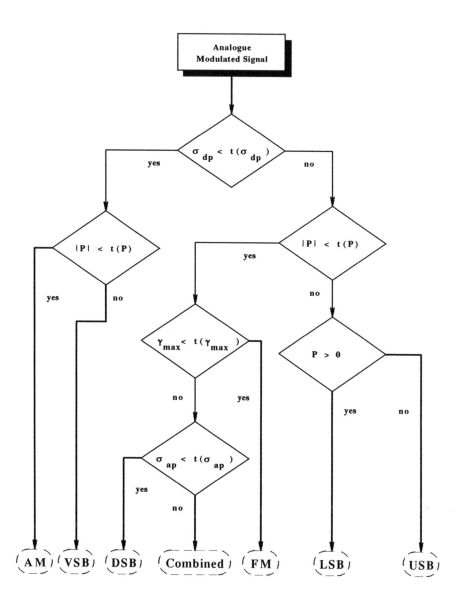

Figure 2.3:   Functional flowchart for AMRA I.

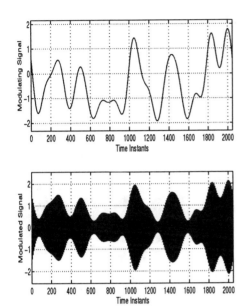

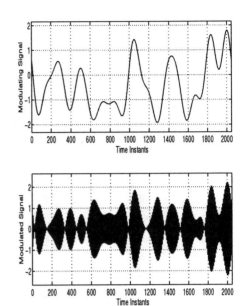

Figure 2.4: Amplitude modulation

Figure 2.5: DSB modulation

Key feature	Corresponding value at ∞ SNR
γ_{max}	59.75
σ_{ap}	0.16
σ_{dp}	0.16
P	0.00

Key feature	Corresponding value at ∞ SNR
γ_{max}	46.20
σ_{ap}	0.30
σ_{dp}	1.45
P	0.00

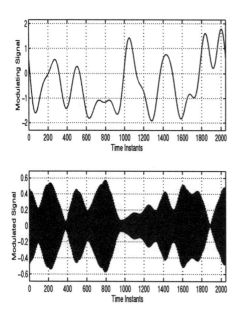

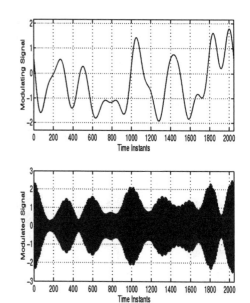

Figure 2.6: VSB modulation

Figure 2.7: LSB modulation

Key feature	Corresponding value at ∞ SNR
γ_{max}	24.98
σ_{ap}	0.24
σ_{dp}	0.24
P	- 0.88

Key feature	Corresponding value at ∞ SNR
γ_{max}	44.35
σ_{ap}	1.45
σ_{dp}	2.88
P	+1.00

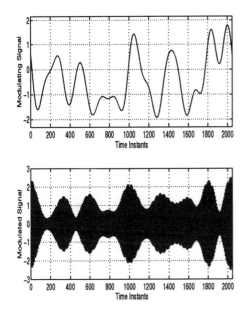

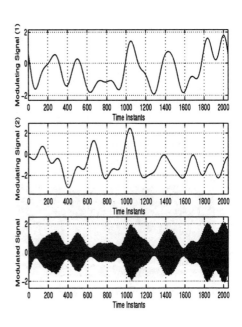

Figure 2.8: USB modulation

Figure 2.9: Combined modulation

Key feature	Corresponding value at ∞ SNR
γ_{max}	44.35
σ_{ap}	1.09
σ_{dp}	2.00
P	-1.00

Key feature	Corresponding value at ∞ SNR
γ_{max}	59.75
σ_{ap}	3.682
σ_{dp}	5.997
P	+ 0.08

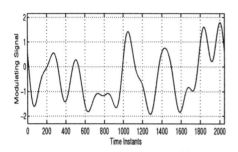

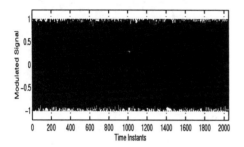

Key feature	Corresponding value at ∞ SNR
γ_{max}	0.00
σ_{ap}	4.01
σ_{dp}	6.53
P	$+\,0.08$

Figure 2.10: Frequency modulation

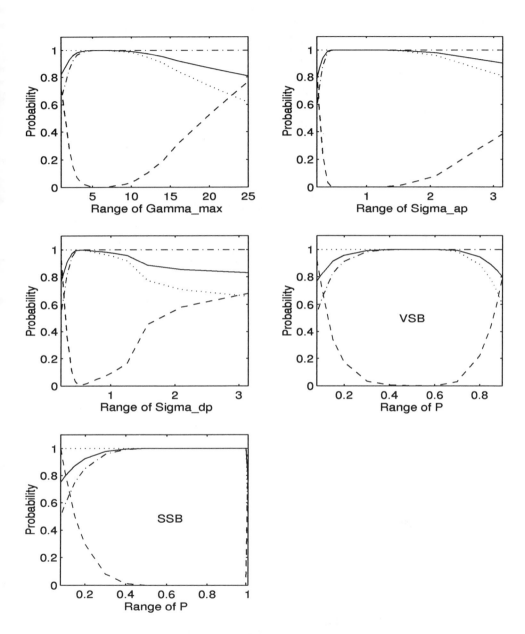

Figure 2.11: Key features threshold determinations for the AMRA I.

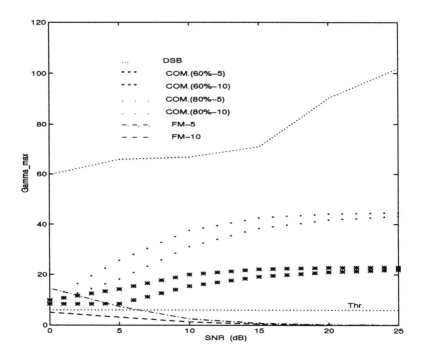

Figure 2.12: Dependence of γ_{max} on the SNR for the AMRA I.

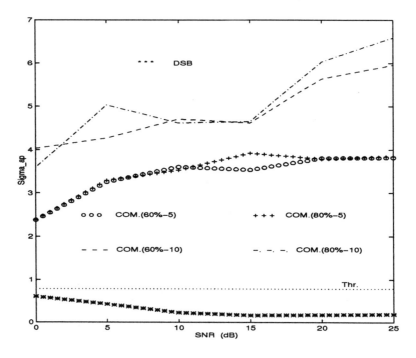

Figure 2.13: Dependence of σ_{ap} on the SNR for the AMRA I.

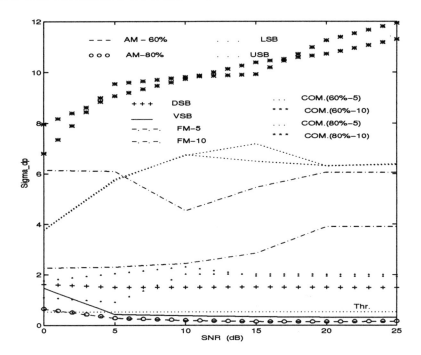

Figure 2.14: Dependence of σ_{dp} on the SNR for the AMRA I.

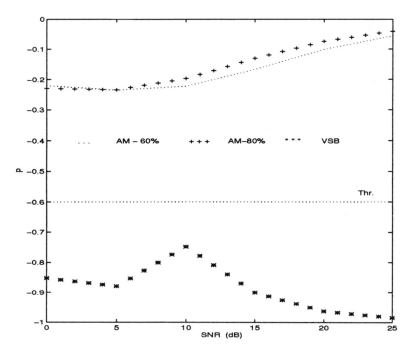

Figure 2.15: Dependence of the ratio P (VSB) on the SNR for the AMRA I.

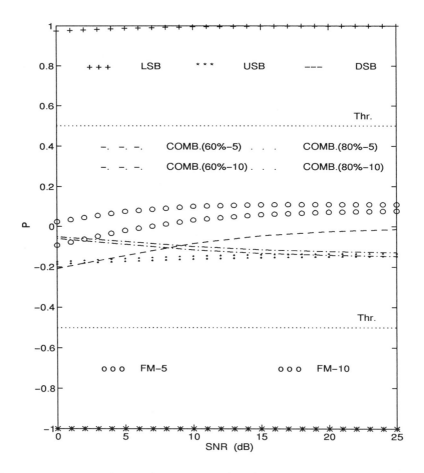

Figure 2.16: Dependence of the ratio P (SSB) on the SNR for the AMRA I.

Chapter 3

Recognition of Digital Modulations

3.1 Introduction

In modern communication systems, digital modulation techniques rather than analogue ones are frequently used. So, the new trend is the digital modulation recognisers.
Most of the digital modulation recognisers discussed in Chapter 1 utilise the pattern
recognition approach such as [4], and [14]. So, they require long signal duration and
the processing time may be very long; and this leads to the use of these algorithms in
off-line analysis. Furthermore, some of these recognisers such as [4] require excessive
computer storage to ensure correct modulation recognition. Indeed, most of them are
assigned to a subset of modulation types of interest. Also, the practical implementation for some of these recognisers such as [14], [21], and [22] is excessively complex.
However, the work on some of these recognisers attempts to identify digital modulations with number of levels > 4. On the other hand, the number of samples used in
the algorithms presented in this thesis to decide about the modulation type is 2048
(equivalent to 1.707 msec.) and this is likely to be suitable for on-line analysis.

This chapter is concerned with introducing three algorithms for digital modulation
recognition, based on the decision-theoretic approach. These algorithms use the same
key features with different time-ordering in the classification algorithm. The modulation types that can be classified with the algorithms developed for digital modulation
recognition are: band-limited ASK2, band-limited ASK4, band-limited PSK2, band-
limited PSK4, band-limited FSK2, and band-limited FSK4 signals. The key features
used in these algorithms are new and they are calculated using the conventional signal

processing tools. Thus, these algorithms can be implemented at extremely low cost and as well they can be used in on-line analysis.

In the next section, a review for the relevant modulation recognisers is introduced. In Section 3.3, the proposed algorithms for digital modulation recognition as well as the extraction of the key features used are introduced. The details of computer simulations of different types of band-limited digitally modulated signals are presented in Section 3.4. The threshold determination and the performance evaluations of the proposed algorithms are introduced in Section 3.5. Finally, the chapter is concluded in Section 3.6.

3.2   Relevant Previous Work

Liedtke is one of the first authors to publish about the modulation recognition process. He also was the first to present the concept of modulation recognition applied to digital modulations. *Liedtke* [4] introduced a modulation recogniser for some types of digital modulations - ASK2, FSK2, PSK2, PSK4, PSK8 and CW. This recogniser utilises the universal demodulator technique. The key features used to discriminate between these types are the amplitude histogram, the frequency histogram, the phase difference histogram, the amplitude variance, and the frequency variance. The classification procedure as shown in Fig. 1.3 comprises the following steps: 1) approximate signal bandwidth estimation, 2) signal demodulation and parameters extraction, 3) statistical computation, and 4) automation of modulation classification. In [4], it is clear that the hardware implementation of this recogniser is excessively complex. In [4], it is claimed that an error free signal, i.e. all the signal parameters are exactly known, can be recognised at SNR $\geq$ 18 dB. In this recogniser, all the analogue modulated signals are classified as noise.

In [9] *DeSimio and Glenn* introduced an adaptive technique for classifying some types of digital modulations - ASK2, PSK2, PSK4 and FSK2. In this recogniser a set of key features derived from the signal envelope, the signal spectra, the signal squared and the fourth power of the signal are used to decide about the modulation type of the intercepted signal. These key features are the mean and variance of the envelope, the magnitude and location of the two largest peaks in the signal spectrum, the magnitude of the spectral component at twice the carrier frequency of the signal square, and the magnitude of the spectral component at four times the carrier frequency of the signal raised to the fourth power. In [9], the classification procedure consists of the following steps: 1) feature vectors extraction, 2) weight vectors generation for each signal class and 3) modulation classification. In [9], the decision functions used are generated using an adaptive technique based on the LMS algorithm. Furthermore, the decision rule used is similar to that applied in the pattern recognition algorithms. So, any intercepting signal is divided into two sets: a training set, which is used to perform the weight vectors and a test set that is used in the decision about the modulation type using the weight vectors generated from the training set. This classifier is trained using the values of the extracted key features at 20 dB SNR. The only thing that mentioned about the performance evaluation of this recogniser is its ability to discriminate between PSK2 and PSK4 at a SNR of 5 dB.

Polydoros and Kim [13] introduced a modulation recogniser, following the decision-theoretic approach, to discriminate between PSK2 and PSK4. In [13], all signal parameters such as the carrier frequency, the initial phase, the symbol rate and the SNR are assumed to be available. This recogniser uses the log-likelihood ratio to estimate the number of levels, M, of the MPSK signals. Also, a comparison between three classifiers for MPSK signals had been introduced. These classifiers are: 1) phase-based classifier (PBC) that is based on the phase difference histogram [4], 2) square-law classifier (SLC) that is based on the fact that squaring of MPSK signal is another MPSK with M/2 phase states, and 3) quasi-log-likelihood ratio (QLLR) classifier which uses the likelihood ratio estimation principles. In [13], it is proved analytically that the performance of the QLLR classifier is significantly better than the PBC or the conventional SLC. Also, it is claimed that the proposed recogniser (QLLR) can be extended to address MPSK signals classification with M > 4.

Hsue and Soliman [14] introduced a modulation recogniser based only on the zero-crossings characteristic of the intercepted signals. The modulation classification procedure as shown in Fig. 1.4 comprises three steps: 1) extraction of the zero-crossing sequence, the zero-crossing interval sequence and the zero- crossing interval difference sequence, 2) inter-symbol transition (IST) detection as well as carrier frequency estimation and 3) decision about the modulation type. The phase and frequency information are derived from the zero-crossing sequence, $\{x(i)\}$, the zero-crossing difference sequence, $\{y(i)\}$, and the zero-crossing interval difference sequence, $\{z(i)\}$. The decision about the modulation type is based on the variance of the zero-crossing interval sequence, G, as well as the frequency and phase difference histograms. This recogniser can be implemented using a parallel processing technique to increase the speed of computations. Also, three processors are recommended, each processor associated with one of the above mentioned three sequences. However, this recogniser is used to report the modulation type of constant amplitude signals such as CW, MPSK, and MFSK. In this recogniser, the classification strategy as shown in Fig. 1.4 comprises two main steps: firstly discrimination of single-tone (CW and MPSK) from multi-tone (MFSK) signals, and secondly determination of the number of states (M). In [14], the discrimination between the single-tone and the multi-tone signals is based on comparing the variance of the zero-crossing difference sequence in the non- weak intervals of a signal with a suitable threshold. The determination of the number of states in single-tone signals is achieved by measuring the similarity of the normalised phase difference histogram. The determination of the number of states in multi-tone signals is based on the number of hills in the zero-crossing interval difference histogram. Finally, the performance of this recogniser was derived from 100 realizations for each modulation type of interest. From the simulation results it is claimed that a reasonable average probability of correct classification is achievable for SNR $\geq$ 15 dB. Anyway, this recogniser cannot identify analogue modulated signals, MASK signals and signals having both amplitude and phase information.

Also, *Soliman and Hsue* [17] introduced another modulation recogniser based on the statistical moments of the intercepted signal phase. In this recogniser, the even order moments of the signal phase are used to estimate the number of levels, M, in MPSK signals. The classification procedure comprises the following steps: 1) instantaneous phase extraction, 2) even order moments computation, 3) thresholds comparison and 4) decision about the modulation type. In [17], all the signal parameters are assumed to be exactly known. Under this assumption, it is claimed that the second order moment is sufficient to discriminate the CW from the MPSK signals, and the eighth order moment is adequate to classify the PSK2 signals with reasonable performance at low

SNR. Also, it is claimed that the suggested classifier is better than the PBC and the SLC.

Assaleh et al. [18] proposed a modulation recogniser for some types of digital modulations. The types that can be classified by this recogniser are: CW, PSK2, PSK4, FSK2, and FSK4. The key features used were derived from the averaged spectrum of the instantaneous frequency (average over M_s successive segments). These key features are the mean and the standard deviation of the averaged instantaneous frequency, the mean and the standard deviation of the instantaneous bandwidth, and the height of the spikes in the differential instantaneous frequency. It is claimed that the performance evaluation of this recogniser is derived from 1000 realizations for each modulation type of interest. Also, it was found that the success rate for the different modulation types of interest is > 99.0% at a SNR of 15 dB. As this recogniser uses averaging over M_s successive segments, long signal duration is required and hence this recogniser is mainly suitable for off-line analysis.

Nagy [20] introduced a modulation recogniser for multichannel systems. This recogniser was accomplished by dividing the analysed signal into individual components and each signal component is classified using a single-tone classifier. The types that have been classified by this recogniser are CW, ASK2, PSK2, PSK4, and FSK2. The developed recogniser comprises three steps. First, detect and filter each signal component in the estimated amplitude spectrum, e.g. the FSK2 is considered as two correlated ASK2 signals. Second, compute the differential phase to discriminate between the different types of single-tone signals. Third, correlate all ASK2 signals to detect the FSK2 signals. In [20], the single-tone classifier is composed of two steps: 1) classification of the ASK2 from the CW, PSK2, and PSK4 using the amplitude histogram, and 2) use the phase histogram to discriminate between CW, PSK2, and PSK4 signals. In [20] the performance of the single-tone classifier of the developed recogniser was derived from 100 realizations for each modulation type of interest. Finally, it is claimed that three single-tone types - CW, PSK2 and PSK4 - have been classified with success rate > 98.0% at 10 dB SNR and the ASK2 with success rate 87.0%.

Beidas and Weber [21] proposed a modulation classifier for MFSK signals. This classifier is based on the time-domain Higher-Order Correlations, and it is used to discriminate between the MFSK signals. Two types of MFSK classifiers are presented in [21] - channelised and non-channelised classifiers. The channelised classifier comprises a bank of matched filters, each of them tuned to one of prescribed frequency locations, and a set of successive correlators. In the non-channelised classifier, each signal is divided into three adjacent subbands - lower, middle, and high. Also, three parallel processors, each of them assigned to one of the subbands, are used. In [21], three algorithms for the non-channelised classifier are considered - the first-order correlation based classifier, in which three correlators and three energy processors are used [21; Fig. 7], the second-order correlation based classifier of the first kind, in which six correlators are used [21; Fig. 8], and the second-order correlation based classifier of the second kind, in which six correlators and three energy processors [21; Fig. 9], are used. All these classifiers are based on comparing the log likelihood function with a suitable threshold to decide about the number of levels of MFSK signals. In [21], it is claimed that the non-channelised classifiers are immune to imperfect knowledge of exact frequency locations.

Huang and Polydoros [22] introduced a modulation recogniser for MPSK signals and it is based on the likelihood function of the instantaneous phase. This recogniser utilises the decision-theoretic approach, as the likelihood function of the instantaneous phase is compared with a suitable threshold. This recogniser can be considered as a generalisation of the modulation recogniser introduced in [13] because it can be used for M > 4. The decision about the modulation type (estimating M for MPSK signals) is carried out according to the following rule

$$L_{PSK2}-L_{PSK4} \underset{PSK4}{\overset{PSK2}{\gtrless}} Threshold$$

where L_{PSK2} and L_{PSK4} are the log-likelihood function of the instantaneous phase of PSK2 and PSK4 signals respectively. In [22] it is claimed that the best performance over all the known MPSK classifiers (PBC and SLC) can be obtained from this classifier. Furthermore, the SLC requires 1.9 dB more SNR than the QLLR classifier to discriminate between PSK2 and PSK4 and also more SNR (2.6 dB) to discriminate between PSK4 and PSK8 signals.

Yang and Soliman [24] modify the modulation recogniser, introduced in [17], that uses the statistical moments to estimate M in the MPSK signals. This modification is in the way of approximating the probability distribution function of the instantaneous phase. In [24], the Fourier series expansion is used for the exact phase distribution approximation instead of the Tikhonov probability density function used in [17]. In [24] it is claimed that this modification offers an improvement of 2 dB for 99.0% success rate and simpler computation for the nth order moments than [17]. In both [17] and [24], nothing is mentioned about the performance evaluation of these two recognisers.

3.3 Developed Digitally Modulated Signal Recognition Algorithms (DMRAs)

Similar to the AMRAs introduced in Chapter 2, instead of taking the decision about the modulation type from only one segment, the decision is derived from the available M_s segments of the intercepted signal frame. Thus, the proposed procedure for digital modulation recognition comprises two main steps: 1) classification of each segment, in which the proposed key features are extracted and compared with the suitable threshold values and 2) classification of a signal frame, in which the majority logic rule is applied to drive a global decision about the modulation type of the received signal frame. Similar ideas have been published in [35].

3.3.1 Classification of each segment

An essential requirement for proposed procedure for digital modulation recognition is to decide in a reliable manner where the intercepted signal comprises the information: in the instantaneous amplitude, in the instantaneous phase, in the instantaneous frequency or in a combination of them. Thus, in the proposed algorithms for digital modulations recognition, the decision about the modulation type is determined by knowing the place of the information content (where the intercepted signal comprises the useful information). In a similar way to that used in the analogue modulations recognition algorithms (see Chapter 2), the suggested procedure for discrimination between the different types of digital modulations based on each segment requires key feature extraction and modulation classification.

Key feature extraction

In the proposed DMRAs, the key features used are derived from three important qualifying parameters - the instantaneous amplitude, the instantaneous phase, and the instantaneous frequency - of the signal under consideration. For the mathematical expressions of these parameters, see Section 1.3.2.

The first key feature is γ_{max}, which is already used in the AMRAs and it is defined by the eqn. (2.1). The second key feature is σ_{ap}, which is also used in the AMRAs and it is defined by the eqn. (2.4). The third key feature is σ_{dp}, which is also used in the AMRAs and it is defined by the eqn. (2.5).

The fourth key feature, σ_{aa}, is defined by:

$$\sigma_{aa} = \sqrt{\frac{1}{N_s} \left(\sum_{i=1}^{N_s} a_{cn}^2(i) \right) - \left(\frac{1}{N_s} \sum_{i=1}^{N_s} |a_{cn}(i)| \right)^2} \tag{3.1}$$

Thus σ_{aa} is the standard deviation of the absolute value of the normalised- centred instantaneous amplitude of a signal segment.

The fifth key feature, σ_{af}, is defined by:

$$\sigma_{af} = \sqrt{\frac{1}{C} \left(\sum_{a_n(i)>a_t} f_N^2(i) \right) - \left(\frac{1}{C} \sum_{a_n(i)>a_t} |f_N(i)| \right)^2} \tag{3.2}$$

where

$$f_N(i) = f_m(i)/r_s, \quad f_m(i) = f(i) - m_f; \quad m_f = \frac{1}{N_s} \sum_{i=1}^{N_s} f(i), \tag{3.3}$$

and r_s is the symbol rate of the digital symbol sequence (in binary modulations, the symbol rate is equal to the bit rate). Thus σ_{af} is standard deviation of the absolute value of the normalised-centred instantaneous frequency, evaluated over the non-weak intervals of a signal segment. It is worth noting that there are many methods for the normalisation such as: 1) normalisation with respect to the maximum value of the instantaneous frequency but some types have zero instantaneous frequency such as MASK signals, so this method is not suitable; 2) normalisation with respect to the mean value of the instantaneous frequency but also, some types have zero mean instantaneous frequency such as MASK and FSK2 signals if the number of ones and zeros are equal, so this method also cannot be used; and 3) normalisation with respect to any signal parameter such as symbol rate, carrier frequency, ..., etc. The author choose the normalisation with respect to the symbol rate as shown in (3.3). Furthermore, for correct recovery of the transmitted message in the digital modulations, it might be necessary to know the symbol rate. Thus, three methods for the symbol duration, T_s ($=\frac{1}{r_s}$), estimation - the level-crossing method, the derivative method, and the wavelet transform method - are introduced by *Azzouz and Nandi* in [32]. It was found that the probability of estimating the symbol duration, within 1% of its true value, is 98.0% for the first method, 90.0% for the second method, and 99.0% for the third method at the SNR of 10 dB. On the whole the wavelet transform method appears to perform better. Furthermore, these three methods are successful in estimating the symbol duration for two and four levels of digital modulations. They can be extended for more than four levels, especially the wavelet transform method. As well they can be used in radar applications - pulse width and pulse repetition frequency measurements. Due to the simplicity of these methods, they can be used for on-line analysis. The details of these three methods are presented by *Azzouz and Nandi* in [36].

A detailed pictorial representation for key feature extraction from a signal segment is shown in Fig. 3.1 in the form of a flowchart.

Modulation classification procedure

Based on the above mentioned five key features, many algorithms can be developed according to the time sequence of applying these key features in the classification algorithm. In this chapter, three algorithms for digital modulations recognition as shown in Figs. 3.2, C.5, and C.6, based on the same key features, are considered. The details of only one algorithm are presented in this chapter. Furthermore, the threshold determinations and the performance evaluations of this algorithm at the SNR of 10 dB, 15 dB and 20 dB are introduced. Meanwhile, the other two algorithms along with their performance evaluations are introduced in Appendix C.2. Moreover, the overall success rates of the three algorithms for digital modulation recognition are presented in this chapter.

DMRA I

In this algorithm, the choice of the γ_{max}, σ_{ap}, σ_{dp}, σ_{aa} and σ_{af}, as key features for digital modulation recognition is based on the following facts:

- γ_{max} is used to discriminate between FSK2 and FSK4 as a subset and ASK2, ASK4, PSK2, and PSK4 as the second subset. As the FSK2, and FSK4 signals have constant instantaneous amplitude, their normalised-centred instantaneous amplitude is zero. Thus, their spectral power densities are zero; i.e. they have no amplitude information ($\gamma_{max} < t(\gamma_{max})$) by their nature. On the other hand, the ASK2 and ASK4 have amplitude information ($\gamma_{max} \geq t(\gamma_{max})$). Furthermore, the PSK2 and PSK4 signals have amplitude information because the band-limitation imposes amplitude information especially at the transitions between successive symbols. So this key feature can be used to discriminate between the signals that have amplitude information (ASK2, ASK4, PSK2, and PSK4) and those do not have amplitude information (FSK2 and FSK4).

- σ_{ap} is used to discriminate between ASK2, ASK4, and PSK2 signals as a subset and the PSK4 signal as the second subset. ASK2 and ASK4 signals have no absolute phase information ($\sigma_{ap} < t(\sigma_{ap})$) by their nature. Also, it is well known that the direct phase of PSK2, similar to the DSB one, takes on values of 0 and π, so its absolute value after centring is constant ($= \pi/2$) such that it has no absolute phase information ($\sigma_{ap} < t(\sigma_{ap})$). On the other hand, the PSK4 has absolute and direct phase information ($\sigma_{ap} \geq t(\sigma_{ap})$) by its natures. So, σ_{ap} can be used to discriminate between the type that has absolute phase information (PSK4) and the types that have no absolute phase information (ASK2, ASK4, and PSK2).

- σ_{dp} is used to discriminate between ASK2, and ASK4 signals as a subset and PSK2 signal as the second subset. ASK2 and ASK4 have no direct phase information ($\sigma_{dp} < t(\sigma_{dp})$) but, the PSK2 signals have direct phase information ($\sigma_{dp} \geq t(\sigma_{dp})$) (the instantaneous phase takes on values of 0 and π). So, σ_{dp} can be used to discriminate between the type that has direct phase information (PSK2) and the types that have no direct phase information (ASK2 and ASK2).

- σ_{aa} is used to discriminate between ASK2 and ASK4 because the normalised-centred instantaneous amplitude of ASK2 signal changes between two levels, equal in magnitude and opposite in the sign, so its absolute value is constant so it has no absolute amplitude information ($\sigma_{aa} < t(\sigma_{aa})$). On the other hand the ASK4 signal has absolute and direct amplitude information ($\sigma_{aa} \geq t(\sigma_{aa})$) by its nature. So, σ_{aa} can be used to discriminate between ASK2 and ASK4 signals.

- σ_{af} is used to discriminate between FSK2 and FSK4 because the normalised-centred instantaneous frequency of FSK2 is changed between two levels, equal in magnitude and opposite in the sign, so its absolute value is constant so it has no absolute frequency information ($\sigma_{af} < t(\sigma_{af})$). On the other hand the FSK4 signal has absolute and direct frequency information ($\sigma_{af} \geq t(\sigma_{af})$) by its nature. So, σ_{af} can be used to discriminate between FSK2 and FSK4 signals.

A detailed pictorial representation of the proposed digital modulation classification procedure is shown in Fig. 3.2 in the form of a flowchart.

3.3.2 Classification of a signal frame

Based on the M_s decisions obtained from the available M_s segments and applying the same procedure used in the AMRAs (Section 2.2), it is possible to obtain a global decision about the modulation type of a signal frame. Due to the simplicity of both the key features extraction as shown in Fig. 3.1 (using the conventional signal processing tools only) and the decision rules as shown in Figs. 3.2, D.5, and D.6, the proposed decision-theoretic DMRAs can be used for on-line analysis.

3.4 Computer Simulations

In this section the software generation of different types of band-limited digitally modulated signals corrupted with band-limited Gaussian noise is introduced and analysed. The simulated band-limited digitally modulated signals and the band-limited Gaussian noise (see Section 2.3) are used in measuring the performance of the developed DMRAs. To increase the degree of realism of the simulated band-limited digitally modulated signals, random pattern sequences instead of reversal pattern sequences are used as modulating signals and the band limitation is carried out according to the usual implementation in practice as is explained later in this section.

3.4.1 Digitally modulated signal simulations

The carrier frequency, f_c , the sampling rate, f_s, and the symbol rate, r_s, were assigned the values 150 kHz, 1200 kHz, and 12.5 kHz respectively. The modulating digital symbol sequence was derived from the modulating speech signal used in the analogue modulated signals simulations (see Section 2.3) in order to make both analogue and digitally modulated generated realizations have almost the same quality with respect to the modulation recognition in presence of a common noise sequence. Thus, the generation of the modulating symbol sequence comprised the following steps:

1. a number $(= 2^{(M-1)})$ of threshold values for the simulated speech signal sequence $\{x(i)\}$, used as a modulating signal for the analogue modulated signal simulations, was calculated;

2. the number of samples per symbol duration, $N_b = 96$, was determined from the assumed symbol and sampling rates; and

3. the simulated speech signal sequence $\{x(i)\}$ was divided into adjacent sets of N_b = 96 samples. Then in every set if the number of samples exceeding a certain threshold, the set was represented by a certain level (e.g. for binary case, if the number of samples exceeding the average value was larger than 47 samples, the set was represented by binary bit "1", otherwise, the set was represented by a binary "0").

MASK, MPSK, and MFSK signals were derived from a general expression

$$s_\theta(i) = a_\theta \cos\left(\frac{2\pi f_\theta i}{f_s} + \phi_\theta\right) ; \quad N_b \geq i \geq 1, \quad \theta = 0, 1, ..., M - 1 \qquad (3.4)$$

Suitable selection of the parameters a_θ, f_θ, and ϕ_θ were made in order to simulate the above mentioned digitally modulated signals ($M = 2$ and 4) and their values for the digital modulations up to 4-levels; i.e., ASK2, ASK4, PSK2, PSK4, FSK2, and FSK4, are presented in Table 3.1.

3.4.2 Band-limiting of simulated modulated signals

It is well known that every communication system has a definite bandwidth. So, any transmitted signal should be band-limited to the pre-defined system bandwidth. Therefore, the simulated digitally modulated signals were band-limited in order to make them represent more realistic test signals for the proposed DMRAs. The band-limitation of the simulated modulated signals was carried out in accordance with the usual implementation in practice. Thus, the band-limitation of digitally modulated signals was exercised after generation. Furthermore, the digital modulation systems are usually implemented in practice as shift-keying systems and hence they cannot be regarded as versions of analogue modulation systems. However, there are two ways of implementing band-limitation of digitally modulated signals. These are: 1) using smoothed square pulses such as raised-cosine square function or Gaussian signal instead of ideal square pulses. Such band-limitation might be similar to that exercised in analogue modulation systems and it is not often used in practice, and 2) shift-keying systems in which the band-limitation is applied on the modulated signal instead. In this case the simulated digitally modulated signals were band-limited to bandwidth, B_w, containing 97.5% of the total average power according to the definition [28]:

$$\int_{f_c-B_w/2}^{f_c+B_w/2} G_s(f)df = 0.975 \int_{-\infty}^{\infty} G_s(f)df \tag{3.5}$$

where $G_s(f)$ is the power spectral density of the digitally modulated signal $s_\theta(t)$.

The analytic expressions of the 97.5 % bandwidth for different types of digital modulations are introduced by *Azzouz and Nandi* in [35]. Furthermore, the theoretical expressions and the simulated values for the bandwidth of the simulated digitally modulated signals are presented in Table 3.2. In our simulations, the value of $f_{mark-\max}$ = $f_c + 2r_s$ for FSK2 as well as for FSK4 if the fourth level exists, otherwise $f_{mark-\max}$ = $f_c + r_s$. Also, $f_{space-\min} = f_c - 2r_s$ for FSK2 and for FSK4 if the first level exists, otherwise $f_{space-\min} = f_c - r_s$. Note that the situation for MFSK signal, with $f_{mark-\max}$ = $f_c + r_s$ and $f_{space-\min} = f_c - r_s$ does not exist. So, MFSK bandwidth is either 7 r_s or 8 r_s.

Typical examples of the modulating signals, modulated signals and the values of all the aforementioned five key features for each modulation type of interest based on 1.707 msec. duration signal (equivalent to 2048 samples) are shown in Figs. 3.3 - 3.8.

3.5 Threshold Determinations and Performance Evaluations

3.5.1 Determination of the relevant thresholds

The implementation of the proposed algorithms for digital modulation recognition, requires the determination of five key features thresholds: $t(\gamma_{\max})$, $t(\sigma_{ap})$, $t(\sigma_{dp})$, $t(\sigma_{aa})$, and $t(\sigma_{af})$ in addition to the normalised-amplitude threshold $a_{t_{opt}}$. Similar to the AMRAs introduced in Chapter 2 and Appendix C.1, the optimum thresholds values are derived from 400 realizations for each modulation type at the SNR of 10dB and 20 dB.

Determination of the optimum key features thresholds values

The aforementioned key features thresholds have been determined based on 400 realizations for each modulation type of interest (ASK2, ASK4, PSK2, PSK4, FSK2, and FSK4) at the SNR of 10 dB and 20 dB. Applying the same procedure presented in Section 2.4, the optimum values for the key feature thresholds, $t(\gamma_{max})$, $t(\sigma_{ap})$, $t(\sigma_{dp})$, $t(\sigma_{aa})$, and $t(\sigma_{af})$ and the corresponding average probability of correct decisions, based on the 400 realizations for the six digitally modulated signals at the SNR of 10 dB and 20 dB, are as shown in Table 3.3 and they can be extracted from Fig. 3.9. It is clear from Table 3.3 and Fig. 3.9 that the optimum threshold values for some key features sometimes take regions instead a single value. In the performance evaluation of the developed algorithms, only one definite value for each key feature threshold is used.

The results of the dependence of the aforementioned key features on SNR for only one realization of different digital modulation types of interest, based on the rule (2.22), for the DMRA I are shown in Figs. 3.10 - 3.14. The optimum key features thresholds values, $t(\gamma_{\max})$, $t(\sigma_{ap})$, $t(\sigma_{dp})$, $t(\sigma_{aa})$, and $t(\sigma_{af})$ are chosen to be 4, $\pi/5.5$, $\pi/5$, 0.25, and 0.4 respectively. These values are chosen to measure the performance of the DMRA I. Furthermore, sample results at the SNR 10 dB, 15 dB and 20 dB are presented in this section. Moreover, the overall success rates are introduced for the developed three DMRAs.

A- Dependence of $\gamma_{\max}$ on SNR

The dependence of γ_{max} on the SNR was computed for six digitally modulated signals - ASK2, ASK4, PSK2, PSK4, FSK2, and FSK4. Of course, only two of these six signals have no amplitude information and they are FSK2 and FSK4 signals. Thus, these six types can be divided into two subsets, A and B. The subset A contains all the types that have amplitude information ($\gamma_{max} \geq t(\gamma_{max})$). These types are: ASK2, ASK4, PSK2, and PSK4. The subset B contains all the types that have no amplitude information ($\gamma_{max} < t(\gamma_{max})$). These types are: FSK2 and FSK4.

The results for one simulation of six digitally modulated signals are presented in Fig. 3.10. From Fig. 3.10, it is clear that for SNR $\geq$ 9 dB the curves corresponding to FSK2 and FSK4 signals fall below the threshold level $t(\gamma_{max}) = 4$. Furthermore, for SNR > 0 dB the curves corresponding to the other types are above the threshold $t(\gamma_{max}) = 4$.

B- Dependence of σ_{ap} on SNR

The dependence of σ_{ap} on the SNR was computed for four digitally modulated signals - ASK2, ASK4, PSK2 and PSK4. Of course three of these four signals have no absolute phase information and they are ASK2, ASK4 and PSK2 signals. Thus, these four modulation types can be divided into two subsets, A and B. The subset A contains the type that has phase information ($\sigma_{ap} \geq t(\sigma_{ap})$), PSK4 signal. The subset B contains all the types that have no absolute phase information ($\sigma_{ap} < t(\sigma_{ap})$). These types are: ASK2, ASK4, and PSK2.

The results for one simulation of these four digitally modulated signals are presented in Fig. 3.11. From Fig. 3.11, it is clear that for SNR > 0 dB the curves corresponding to ASK2, ASK4, and PSK2 signals fall below the threshold level $t(\sigma_{ap}) = \pi/5.5$. Furthermore, for SNR > 0 dB the curve corresponding to the PSK4 signal is above the threshold $t(\sigma_{ap}) = \pi/5.5$.

C- Dependence of σ_{dp} on SNR

The dependence of σ_{dp} on the SNR was computed for three digitally modulated signals - ASK2, ASK4, and PSK2. Of course two of these three signals have no direct phase information and they are ASK2 and ASK4 signals. Thus, these three modulation types can be divided into two subsets, A and B. The subset A contains the type that has direct phase information ($\sigma_{dp} \geq t(\sigma_{dp})$), PSK2 signal. The subset B contains all the types that have no direct phase information ($\sigma_{dp} < t(\sigma_{dp})$) and these types are: ASK2 and ASK4.

The results for one simulation of these three digitally modulated signals are presented in Fig. 3.12. From Fig. 3.12, it is clear that for SNR $>$ 0 dB the curves corresponding to ASK2 and ASK4 signals fall below the threshold level $t(\sigma_{dp}) = \pi/5$. Furthermore, for SNR $>$ 0 dB the curve corresponding to the PSK2 is above the threshold $t(\sigma_{dp}) = \pi/5$.

D- Dependence of σ_{aa} on SNR

The dependence of σ_{aa} on the SNR was computed for two digitally modulated signals - ASK2 and ASK4. The ASK4 signal has absolute amplitude information ($\sigma_{aa} \geq t(\sigma_{aa})$), so it represents the subset A while ASK2 has no absolute amplitude information ($\sigma_{aa} < t(\sigma_{aa})$), so it represents the subset B.

The results for one simulation of these two types are presented in Fig. 3.13. From Fig. 3.13, it is clear that for SNR $>$ 0 dB the curve corresponding to the ASK2 signal falls below the threshold level $t(\sigma_{aa})= 0.25$. Furthermore, for SNR $>$ 0 dB the curve corresponding to ASK4 is above the threshold $t(\sigma_{aa})= 0.25$.

E- Dependence of σ_{af} on SNR

The dependence of σ_{af} on the SNR was computed for two digitally modulated signals - FSK2 and FSK4. The FSK4 signal has absolute instantaneous frequency information ($\sigma_{af} \geq t(\sigma_{af})$), so it represent the subset A while FSK2 has no absolute frequency information ($\sigma_{af} < t(\sigma_{af})$), so it represent the subset B .

The results for one simulation of these two types are presented in Fig. 3.14. From Fig. 3.14, it is clear that for SNR $>$ 7 dB the curve corresponding to FSK2 signal falls below the threshold level $t(\sigma_{af})= 0.4$. Furthermore, for SNR > 0 dB the curve corresponding to FSK4 is above the threshold $t(\sigma_{af})= 0.4$.

Determination of the normalised amplitude threshold $a_{t_{opt}}$

The normalised amplitude threshold $a_{t_{opt}}$ was determined in a similar way to that used in the AMRAs. It was found that the simulated MFSK (M=2 and 4) signals imposed an upper bound equal to **1.05**. Furthermore, it was found that the optimum normalised amplitude threshold is in the range $a_{t_{opt}} = [0.9:1.05]$. In the proposed DMRAs, $a_{t_{opt}}$ is set to be **1**.

3.5.2   Performance Evaluations

The results of the performance evaluations of the proposed procedure for digitally modulated signals recognition, DMRA I, are derived from 400 realizations for each modulation type of interest. The values of the optimum key feature thresholds, $t(\gamma_{max})$, $t(\sigma_{ap})$, $t(\sigma_{dp})$, $t(\sigma_{aa})$, and $t(\sigma_{af})$ are chosen to be 4, $\pi/5.5$, $\pi/5$, 0.25, and 0.4 respectively. The corresponding results are summarised in Tables 3.4, 3.5, and 3.6 for three SNR values - 10 dB, 15 dB and 20 dB. Each of the modulation types at each SNR was simulated 400 times and each realization was 2048 samples. Consider Table 3.4 for example, it can be observed that all the digital modulation types of interest have been correctly classified with more than 98.0% success rate. The results in Table 3.6 corresponds to a SNR of 20 dB. It is clear that the success rate for correct modulation recognition has increased with increasing the SNR, and now all types have been classified with success rate $> 99.0\%$ and four of the six modulation types have been successfully classified every time (100% success rate).

3.5.3   Processing Time and Computational Complexity

The processing time and computational power are measured using the MATLAB software on the SUN SPARC station **20**. The values of the processing time (measured in seconds) and the computational power (measured in Megaflops) measured on the SUN SPARC station **20** are shown in Table 3.7. The processing time and the number of Megaflops, required to take a decision about the modulation type, correspond to only one signal segment in each case. The number in Table 3.7 are the average values of the measurements for 10 different realizations of each modulation type of interest at ∞ SNR.

3.6 Conclusions

The aim of this chapter has been to introduce fast and reliable DMRAs. The current approach has been to carry out this task using the decision-theoretic approach. A number of new key features are proposed to fulfil the requirement of these algorithms. These key features are calculated using conventional signal processing methods. So, the proposed algorithms can be implemented at extremely low cost. The most interesting observation is that these key features can be extended to a larger number of levels ($M > 4$). Furthermore, the decision on the modulation type is achieved in a very short time due to the simplicity of key feature extraction and the decision rules used in the proposed algorithms. The only requirement to use these recognisers for on-line analysis is to finish the determination of the key features thresholds values in the off-line time.

Extensive simulations of six digitally modulated signals have been carried out at different SNR. Sample results have been presented at the SNR of 10 dB, 15 dB and 20 dB only. It is found that the threshold SNR for successful modulation recognition (at a success rate $> 98.0\%$) is about 10 dB, which is an improvement in the SNR threshold over previous results reported in [4], [14], and [18]. Also presented are measured processing times and required computational power of the proposed algorithms for different types of digital modulations considered in this thesis.

In the developed algorithms, the key feature threshold values are derived from 400 realizations of each modulation type of interest at SNR of 10 dB and 20 dB. The overall probability of correct decision at three SNR values - 10 dB, 15 dB and 20 dB - is shown in Table 3.8. It is worth noting that the three algorithms achieve the same success rates for different modulation types at the SNR of 10 dB, 15 dB and 20 dB, but the performance of the DMRA III is slightly better at the SNR of 20 dB.

Modulation Type	M	a_θ	$f_\theta[kHz]$	ϕ_θ
MASK	2	$0.8\,\theta + 0.2$	f_c	0
	4	$0.25\,\theta + 0.25$	f_c	0
MPSK	2	1	f_c	$(1-\theta)\pi$
	4	1	f_c	$\theta\frac{\pi}{2}$
MFSK	2	1	$4r_s\theta + f_c - 2r_s$	0
	4	1	$f_c - (\theta+1)r_s$ if $\theta < 2$ $f_c + (\theta-1)r_s$ if $\theta \geq 2$	0

Table 3.1: Digitally modulated signals parameter selection.

Modulation Type	Theoretical expression	Simulated value (kHz)
MASK	$4\,r_s$	50
MPSK	$6\,r_s$	75
MFSK	$[\mid f_{mark-\max} - f_{space-\min} \mid +4r_s]$	100 or 87.5

Table 3.2: Bandwidths of digitally modulated signals.

Key Features Thresholds	Optimum Value	$P_{av}(x_{opt})$
$t(\gamma_{max})$	4	99.6%
$t(\sigma_{ap})$	$\pi/5.5$	100%
$t(\sigma_{dp})$	$[\pi/6.5\text{-}\pi/2.5]$	100%
$t(\sigma_{aa})$	0.25	99.6%
$t(\sigma_{af})$	0.4	99.8%

Table 3.3: Optimum key features threshold values for the DMRA I.

Simulated	Deduced Modulation Type					
Modulation Type	ASK2	ASK4	PSK2	PSK4	FSK2	FSK4
ASK2	98.3%	1.7%	-	-	-	-
ASK4	-	100%	-	-	-	-
PSK2	-	-	99.3%	-	0.7%	-
PSK4	-	-	-	98.8%	1.2%	-
FSK2	-	-	-	0.5%	99.5%	-
FSK4	-	-	-	1.0%	0.7%	98.3%

Table 3.4: Confusion matrix for the DMRA I [based on 400 realizations] at SNR = 10 dB.

Simulated	Deduced Modulation Type					
Modulation Type	ASK2	ASK4	PSK2	PSK4	FSK2	FSK4
ASK2	98.3%	1.7%	-	-	-	-
ASK4	0.2%	99.8%	-	-	-	-
PSK2	-	-	99.3%	-	0.7%	-
PSK4	-	-	-	98.8%	1.2%	-
FSK2	-	-	-	0.5%	99.5%	-
FSK4	-	-	-	1.0%	0.5%	98.5%

Table 3.5: Confusion matrix for the DMRA I [based on 400 realizations] at SNR = 15 dB.

Simulated	Deduced Modulation Type					
Modulation Type	ASK2	ASK4	PSK2	PSK4	FSK2	FSK4
ASK2	100%	-	-	-	-	-
ASK4	-	100%	-	-	-	-
PSK2	-	-	99.3%	-	0.7%	-
PSK4	-	-	-	99.8%	0.2%	-
FSK2	-	-	-	-	100%	-
FSK4	-	-	-	-	-	100%

Table 3.6: Confusion matrix for the DMRA I [based on 400 realizations] at SNR = 20 dB.

Modulation Type	Computational power (Megaflops)	Processing time (Sec.)
ASK2	0.45	6.51
ASK4	0.45	6.55
PSK2	0.43	6.32
PSK4	0.43	6.14
FSK2	0.49	10.77
FSK4	0.49	10.71

Table 3.7: Measured computational power and processing times of the DMRA I on SPARC 20.

SNR (dB)	Overall success rates	
	I & II	III
10	99.0%	99.0%
15	99.0%	99.0%
20	99.9%	100%

Table 3.8: Overall success rates for the developed DMRAs.

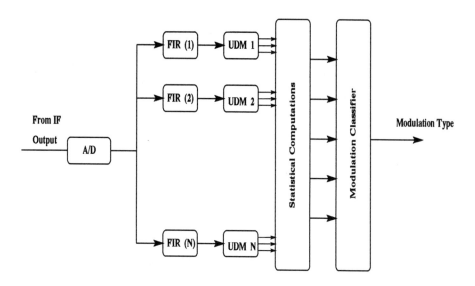

Figure 3.1: Classification procedure for *Liedtke* modulation recogniser [4].

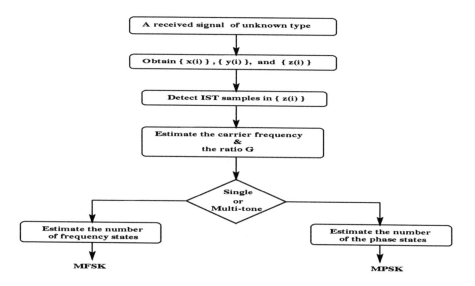

Figure 3.2: Simplified block scheme of a recogniser based on the zero-crossings of a signal [14].

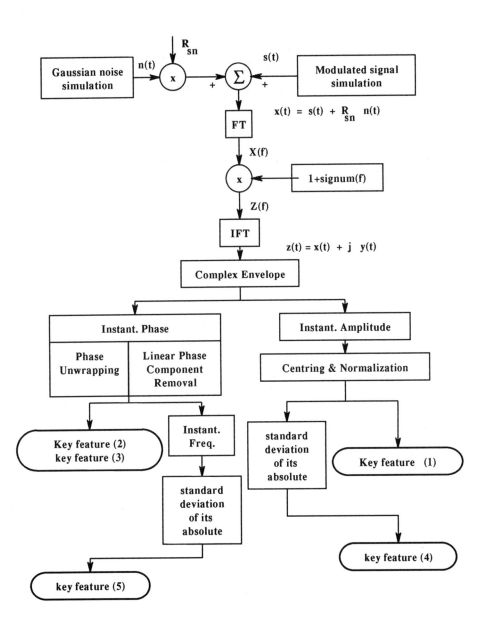

Figure 3.3: Functional flowchart for key features extraction in the DMRAs.

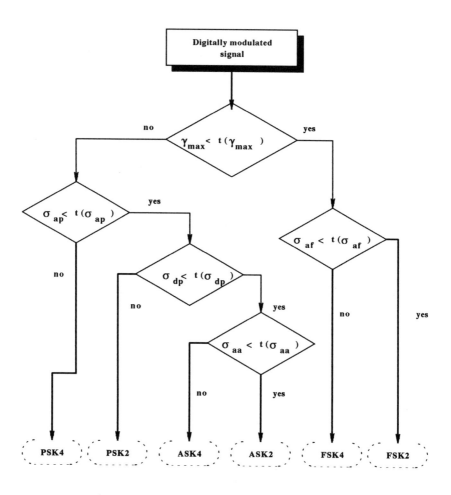

Figure 3.4: Functional flowchart for the DMRA I.

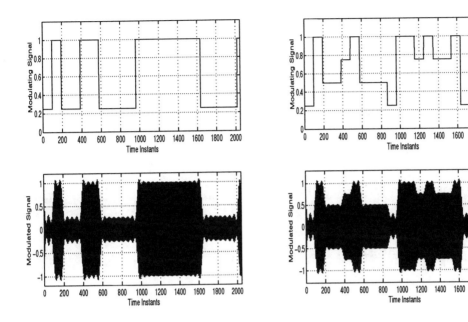

Figure 3.5: B.L. ASK2 modulation   Figure 3.6: B.L. ASK4 modulation

Key feature	Corresponding value at ∞ SNR
γ_{max}	121.35
σ_{ap}	0.03
σ_{dp}	0.03
σ_{aa}	0.00
σ_{af}	0.00

Key feature	Corresponding value at ∞ SNR
γ_{max}	42.54
σ_{ap}	0.03
σ_{dp}	0.03
σ_{aa}	0.32
σ_{af}	0.000

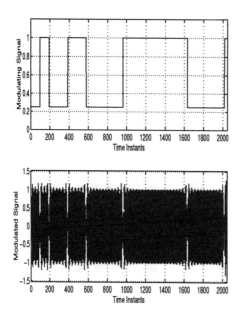

Figure 3.7: B.L. PSK2 modulation.

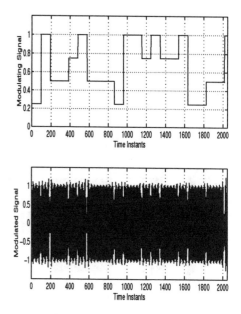

Figure 3.8: B.L. PSK4 modulation.

Key feature	Corresponding value at ∞ SNR
γ_{max}	6.15
σ_{ap}	0.304
σ_{dp}	1.57
σ_{aa}	0.00
σ_{af}	0.10

Key feature	Corresponding value at ∞ SNR
γ_{max}	6.85
σ_{ap}	4.77
σ_{dp}	6.67
σ_{aa}	0.00
σ_{af}	0.13

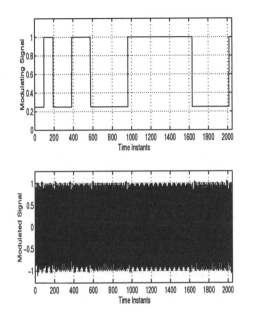

Figure 3.9: B.L. FSK2 modulation.

Figure 3.10: B.L. FSK4 modulation.

Key feature	Corresponding value at ∞ SNR
γ_{max}	0.15
σ_{ap}	6.39
σ_{dp}	9.47
σ_{aa}	0.00
σ_{af}	0.06

Key feature	Corresponding value at ∞ SNR
γ_{max}	0.09
σ_{ap}	5.62
σ_{dp}	8.50
σ_{aa}	0.00
σ_{af}	0.48

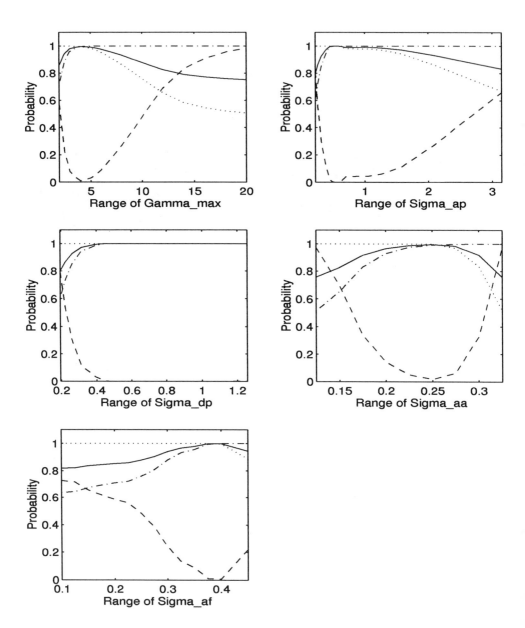

Figure 3.11: Key features thresholds determinations for the DMRA I.

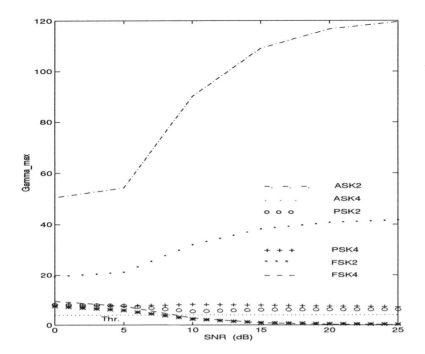

Figure 3.12: Dependence of γ_{max} on the SNR for the DMRA I.

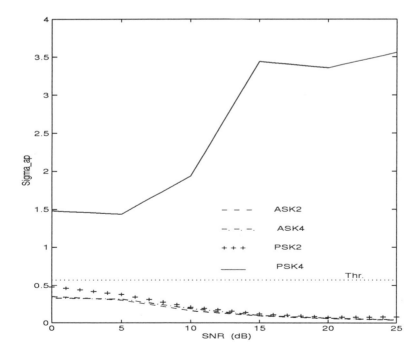

Figure 3.13: Dependence of σ_{ap} on the SNR for the DMRA I.

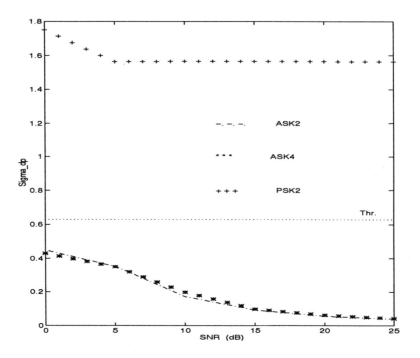

Figure 3.14: Dependence of σ_{dp} on the SNR for the DMRA I.

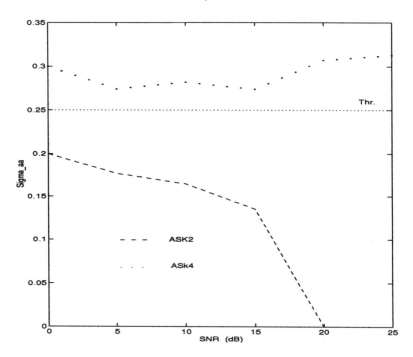

Figure 3.15: Dependence of σ_{aa} on the SNR for the DMRA I.

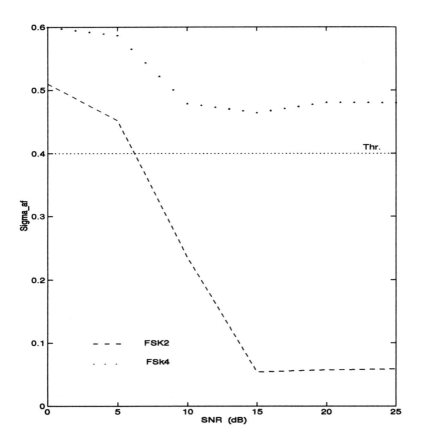

Figure 3.16: Dependence of σ_{af} on the SNR for the DMRA I.

Chapter 4

Recognition of Analogue & Digital Modulations

4.1 Introduction

Some of the modulation recognisers for both analogue and digital modulations discussed in Chapter 1 utilise the pattern recognition approach, which requires long signal duration and strong SNR to decide about the modulation type of an RF signal. So, these recognisers are mainly used in off-line analysis. Furthermore, none of these recognisers considered VSB and combined modulation signals. In this chapter a set of fast algorithms for the recognition of both analogue and digital modulations without any a priori information about the nature of a signal, whether it is analogue or digital, is introduced. Moreover, the AMRAs introduced in Chapter 2 and Appendix C.1 are concerned with only analogue modulated signals. Also, the DMRAs introduced in Chapter 3 and Appendix C.2 are concerned with digitally modulated signals only. So, for these algorithms, there is a priori information about the nature of the signal under consideration. Sometimes, there is no a priori information about the nature of a signal. In this case the modulation recognition algorithms introduced in Chapter 2, Chapter 3, Appendix C.1, and Appendix C.2 cannot discriminate between some types of modulations such as MASK and AM, MFSK and FM, PSK2 and DSB, and PSK4 and combined modulated signals. It is worth noting that the only difference between such modulation types is in the nature of the modulating signal used to generate these modulation types. In the analogue modulated signals a speech signal is used as modulating signal while a digital symbol sequence is used as a modulating signal in the

108

digitally modulated signals.

In this chapter, three algorithms for both analogue and digital modulation recognition, based on the decision-theoretic approach, use the same key features as introduced. The key features used are new and they are extracted using the conventional signal processing tools. The modulation types that can be classified by these recognisers are all the analogue modulation types that are mentioned in Chapter 2 as well as all the digital modulation types that are mentioned in Chapter 3. Computer simulations of different types of band-limited analogue and digitally modulated signals corrupted by a band-limited Gaussian noise sequence have been carried out to measure the performance of the developed algorithms. It is worth noting that a complete illustration of the analogue modulated signals simulations is introduced in Chapter 2 and of the digitally modulated signals used is presented in Chapter 3. In the next section, a review of the more recent published papers in this area is presented. In Section 4.3, the proposed algorithms for both analogue and digital modulation recognition along with the extraction of the proposed key features are introduced. The thresholds determinations and the performance evaluations of the proposed algorithms are introduced in Section 4.4. Finally, the chapter is concluded in Section 4.5.

4.2 Relevant Previous Work

In [5] *Callaghan et al.* proposed a modulation recogniser utilising the envelope and the zero-crossings characteristics of the intercepted signal. This recogniser uses a phase-locked loop (PLL) for carrier recovery in the weak intervals of the intercepted signal segment. It is worth noting that in some modulation types such as DSB, MPSK and AM with high modulation depth, the carrier frequency may be severely suppressed or absent. Carrier recovery during the suppressed portions (weak intervals) is equivalent to receiving a signal with very low SNR. So, using the PLL for carrier recovery (hardware solution) overcomes the problem of needing high SNR for accurate instantaneous frequency estimation from the zero-crossings. Also, the accuracy of this recogniser deteriorates rapidly if the receiver is not perfectly tuned to the centre frequency. The modulation types that can be recognised by this recogniser are CW, AM, FM, FSK2. In [5], it is claimed that this recogniser requires SNR $\geq$ 20 dB for the correct recognition of the different modulation types of interest. Furthermore, this recogniser cannot discriminate the MPSK and the DSB signals due to the incorrect estimate of the instantaneous frequency of the intercepted signal that results from the effect of the noise on the weak intervals of a signal segment.

Jondral [6] proposed a modulation recogniser utilising the pattern recognition approach for two types of analogue modulated signals - AM and SSB - as well as for four types of digitally modulated signals - ASK2, PSK2, FSK2 and FSK4. The key features used are derived from the instantaneous amplitude, phase and frequency. These key features are the instantaneous amplitude, phase difference and frequency histograms. In this recogniser, the instantaneous amplitude histogram is computed for the normalised instantaneous amplitude. The normalisation is done with respect to the maximum values of the intercepted signal. As this classifier uses the pattern recognition approach, the received signal is divided into two adjacent sets: learning set and test set. The segment length used in this recogniser is 4096 samples for each modulation type. In [6] real signals have been used and it is claimed that all the above mentioned modulation types have been classified with success rate $\geq 90.0\%$ except for SSB (=83.0%) and FSK4 (=88.0%).

Aisbett in [8] developed three new signal parameters, and they are A^2, AA' and $A^2\theta'$ where A is the signal envelope, A' is the signal envelope derivative and θ' is the instantaneous frequency. In [8], it is claimed that the new parameters are more noise resistant compared with the standard parameters - A and θ'. The key features used are the peak and tail values of the developed parameters - A^2, AA' and $A^2\theta'$. The estimation of these key features is derived from 10 realizations, each with 90 msec. length, for each modulation type of interest. Also, another unbiased key feature is added by *Asibett* in [8] and it is defined as the variance of the squared instantaneous amplitude minus its squared mean. The modulation types that can be classified by this recogniser are: AM, DSB, FM, ASK2, PSK2, FSK2, and CW signals. In [8], it is claimed that the success rate of the discrimination between the modulation types appears to be at least good for strong signals. Also, it is mentioned that the performance of the developed recogniser is better than that utilise the standard parameters - A and θ'.

Petrovic et al. [11] suggested a modulation recogniser based on the parameters variations and the zero-crossing rate of the AM detector output as well as the parameters variations in the FM detector output. The modulation types that can be classified by this recogniser are AM, FM, SSB, CW, ASK2, FSK2. This modulation recogniser as shown in Fig. 1.5 comprises three main steps: 1) AM and FM demodulation, 2) key features extraction, and 3) modulation classification. In [11], three key features are derived from the AM detector output and they are: detect the presence of the signal, measure the amplitude variations and the instantaneous amplitude zero-crossing rate. Also, for the FM detector output a narrow band and a wide band FM detection are performed. The only thing mentioned about the performance evaluation is that the results of the preliminary test with real signals show the successfulness of this recogniser.

Martin [12] proposed a modulation recogniser for some analogue and digital modulation types. These types are: AM, FM, SSB, CW, ASK2, and FSK2. The key features used are derived from the instantaneous amplitude, the IF signal spectrum and its derivative. These key features are: the amplitude histogram, the signal bandwidth, and the relationship between the spectral components. In [12] real signals have been used and it is claimed that all the modulation types of interest have been classified with success rate > 90.0% except for FM (=80.0%).

Dominguez et al. [16] introduced a modulation recogniser which is a general approach for both analogue and digital modulations. This recogniser is concerned with some types of analogue and digitally modulated signals. These types are AM, DSB, SSB, FM, CW, noise and the digitally modulated signals up to 4-levels - ASK2, ASK4, PSK2, PSK4, FSK2, and FSK4. This recogniser comprises three subsystems: 1) pre-analysis subsystem, 2) features extraction subsystem and 3) classifier subsystem. The recognition algorithm is based on the histograms of the instantaneous amplitude, phase and frequency. In [16], it is claimed that this recogniser performed well and all the aforementioned modulation types have been correctly classified at SNR ≥ 40 dB. In [16], it is claimed that the performance of the developed algorithm is derived from 100 realizations for each modulation type of interest. At an SNR of 10 dB the probability of correct modulation recognition is 0% for all the digital modulation types except for PSK4 (=7.0%) and at 15 dB SNR the performance is still wanting especially for FSK4 (=56.0%) , FSK2 (=84.0%) and ASK4 (=87.0%). Furthermore, the number of samples per segment used in the performance evaluation of this recogniser is 3000 samples.

However, it should be noted that this work attempts to identify most of the analogue and digital modulation types considered here in this thesis.

4.3 Developed Analogue & Digitally Modulation Recognition Algorithms (ADMRAs)

Similar to the AMRAs introduced in Chapter 2 and Appendix C.1, and the DMRAs introduced in Chapter 3 and Appendix C.2, the same global procedure used in these two groups of algorithms is applied here in the proposed ADMRAs; i.e. the decision is derived from all the available M_s segments of a signal frame. Thus, the proposed global procedure for both analogue and digital modulations recognition comprises two main steps: 1) classification of each segment, and 2) classification of a signal frame. Similar ideas have been published in [37].

4.3.1 Classification of each segment

From every available segment, the suggested procedure to discriminate between the different modulation types requires key features extraction and modulation classification, which gives a decision about the modulation type presents in each segment by comparing each key feature with a suitable threshold.

Key features extraction

All the key features used in the proposed ADMRAs are derived from three important qualifying parameters, except the signal spectrum symmetry which is derived from the RF signal spectrum. These parameters are the instantaneous amplitude, the instantaneous phase and the instantaneous frequency of the intercepted signal. The four key features, used in Chapter 2 for the AMRAs, are also used here for the ADMRAs, and they are:

- the maximum value of the spectral power density of the normalised-centred instantaneous amplitude, γ_{max}, and it is defined by the eqn. (2.1);

- the standard deviation of the absolute value of the centred non-linear component of the instantaneous phase in the non-weak intervals of a signal segment, σ_{ap}, and it is defined by the eqn. (2.4);

- the standard deviation of the direct value of the centred non-linear component of the instantaneous phase in the non-weak intervals of a signal segment, σ_{dp}, and it is defined by the eqn. (2.5); and

- the ratio, P, which measure the spectrum symmetry of the RF signal around its carrier frequency and it is defined by the eqn. (2.6).

Two of the key features used in Chapter 3 for the DMRAs are also used here in the ADMRAs, and they are:

- the standard deviation of the absolute value of the normalised-centred instantaneous amplitude of a signal, σ_{aa}, and it is defined by the eqn. (3.1); and

- the standard deviation of the absolute value of the normalised-centred instantaneous frequency evaluated over the non-weak intervals of a signal segment, σ_{af}, and it is defined by the eqn. (3.2).

To complete the proposed global procedure for analogue and digital modulations recognition, three new key features are introduced. These are:

- the standard deviation of the normalised-centred instantaneous amplitude in the non-weak intervals of a signal segment, defined by

$$
\sigma_a = \sqrt{\frac{1}{C}\left(\sum_{a_n(i)>a_t} a_{cn}^2(i)\right) - \left(\frac{1}{C}\sum_{a_n(i)>a_t} a_{cn}(i)\right)^2}, \qquad (4.1)
$$

- the kurtosis of the normalised-centred instantaneous amplitude, μ_{42}^a, defined by

$$
\mu_{42}^a = \frac{E\left\{a_{cn}^4(t)\right\}}{\left\{E\left\{a_{cn}^2(t)\right\}\right\}^2} \qquad (4.2)
$$

where $a_{cn}(t)$ is the normalised-centred instantaneous amplitude as expressed by the eqn. (2.2),

- the kurtosis of the normalised-centred instantaneous frequency, μ_{42}^f, defined by

$$
\mu_{42}^f = \frac{E\left\{f_N^4(t)\right\}}{\left\{E\left\{f_N^2(t)\right\}\right\}^2} \qquad (4.3)
$$

where $f_N(t)$ is the normalised-centred instantaneous frequency as expressed by the eqn. (3.3).

A detailed pictorial representation for the key features extraction from an RF signal is shown in Fig. 4.1 in the form of a flowchart.

Modulation classification procedure

Based on the aforementioned nine key features, many algorithms can be generated according to the sequence of applying these key features in the classification algorithm. In this chapter, three algorithms for both analogue and digital modulations recognition as shown in Figs. 4.2, C.7 and C.8, using the same key features, are considered. The details of only one algorithm are presented. Furthermore, the performance measures for all the analogue and digital modulation types of interest at the SNR of 15 dB and 20 dB are introduced for that algorithm. Also, the over all success rates at the SNR of 15 dB and 20 dB for different ADMRAs are introduced in this chapter. Meanwhile, the other two algorithms along with their performance evaluations are introduced in Appendix C.3.

ADMRA I

In this algorithm, the choice of γ_{max}, σ_{ap}, σ_{dp} and the ratio P is based on some facts similar to those mentioned in Chapter 2. Also, the choice of σ_{aa} and σ_{af} is based on the same facts discussed in Chapter 3. Furthermore, the choice of σ_a, μ_{42}^a, and μ_{42}^f as key features for the proposed ADMRAs is based on the following facts [38]:

- σ_a is used to discriminate between the DSB and PSK2 signals as well as to discriminate between the combined (AM-FM) and PSK4 signals. The PSK2 and PSK4 signals have no amplitude variations except at the transition between the successive symbols; i.e the normalised-centred instantaneous amplitude is constant $(= 0)$ over the symbol duration $(\sigma_a < t(\sigma_a))$. On the other hand, the DSB and the combined (AM-FM) signals have amplitude information $(\sigma_a > t(\sigma_a))$. So, σ_a can be used to discriminate between the DSB and PSK2 as well as to discriminate between the combined (AM-FM) and PSK4 signals

- μ_{42}^a is used to discriminate between the AM signals as a subset and the MASK signals (ASK2 and ASK4) as the second subset. This key feature as defined by (4.2) is used to measure the "compactness of the instantaneous amplitude distribution". So it can be used to discriminate between the signals, in which the instantaneous amplitude has high compact distribution $(\mu_{42}^a > t(\mu_{42}^a))$ such as the AM signals (related to the speech signal), and those, in which the instantaneous amplitude have less compact distribution ($\mu_{42}^a < t(\mu_{42}^a)$) such as MASK (ASK2 and ASK4) signals (related to the symbol sequence).

- μ_{42}^f is used to discriminate between the FM signals as a subset and the MFSK signals (FSK2 and FSK4) as the second subset. This key features as defined in (4.3) is used to measure the "compactness of the instantaneous frequency distribution". So it can be used to discriminate between the FM signal, in which the instantaneous frequency (related to the speech signal) have high compact distribution ($\mu_{42}^f > t(\mu_{42}^f)$), and MFSK, in which the instantaneous frequency (related to the symbol sequence) have less compact distribution ($\mu_{42}^f < t(\mu_{42}^f)$).

A detailed pictorial representation of the proposed analogue and digital modulations recognition procedure is shown in Fig. 4.2 in the form of a flowchart.

4.3.2 Classification of a signal frame

As it is possible to obtain different classifications of these M_s segments, the majority logic rule is applied ; i.e. select the classification with largest number of repetitions. If two or more classifications have equal maximum numbers of repetitions, they are regarded as candidates for optimal decision. Hence, the global decision about the modulation type of an RF signal is determined in a similar way to that used in either the AMRAs or the DMRAs.

Due to the simplicity of both the key features extraction and the decision rules used, the proposed algorithms for both analogue and digital modulations recognition can be used for on-line analysis. The main requirement to use the proposed algorithms for on-line analysis is that the key feature thresholds be determined off-line before using the algorithms for on-line analysis.

Typical examples of the modulating signals, modulated signals for all types of analogue and digital modulations of interest have been introduced in Chapter 2 and Chapter 3 respectively. Furthermore, the values of the key features required for the ADMRAs only (not required for the other algorithms; analogue only or digital only) - σ_a, μ_{42}^a, and μ_{42}^f - measured for one and the same realization used in Chapter 2 and Chapter3, are shown in Table 4.1.

4.4 Threshold Determinations and Performance Evaluations

4.4.1 Determination of the relevant thresholds

The implementation of the proposed algorithms for both analogue and digital modulations recognition, requires the determination of nine important key feature thresholds: $t(\gamma_{\max})$, $t(\sigma_{ap})$, $t(\sigma_{dp})$, $t(P)$, $t(\sigma_{aa})$, $t(\sigma_{af})$, $t(\sigma_a)$, $t(\mu_{42}^a)$, and $t(\mu_{42}^f)$ in addition to the normalised-amplitude threshold $a_{t_{opt}}$.

Determination of the key feature threshold values

The aforementioned key feature thresholds have been determined based on 400 realizations for each modulation type of interest - the twelve analogue modulation types used in Chapter 2 and the six digital modulation types used in Chapter 3 - at SNR of 15 dB and 20 dB. Applying the same procedure presented in Section 2.4, the optimum values for the key feature thresholds, $t(\gamma_{\max})$, $t(\sigma_{ap})$, $t(\sigma_{dp})$, $t(P)$, $t(\sigma_{aa})$, $t(\sigma_{af})$, $t(\sigma_a)$, $t(\mu_{42}^a)$, and $t(\mu_{42}^f)$, based on the 400 realizations for the eighteen modulated signals are as shown in Table 4.2 and they can be extracted from Fig. 4.3. Anyway, the optimum values for the key feature thresholds and the corresponding average probability of correct decisions, based on the 400 realizations at SNR of 15 dB and 20 dB, are shown in Table 4.2 and they can be extracted from Fig. 4.3.

The optimum values for the key feature thresholds - $t(\gamma_{\max})$, $t(\sigma_{ap})$, $t(\sigma_{dp})$, $t(P)$, $t(\sigma_{aa})$, $t(\sigma_{af})$, $t(\sigma_a)$, $t(\mu_{42}^a)$, and $t(\mu_{42}^f)$ - used to measure the performance of the ADMRA I are chosen to be 2.5, $\pi/5.5$, $\pi/6$, 0.6, 0.25, 0.4, 0.25 (PSK2) and 0.15 (PSK4), 2.15, and 2.03 respectively. The results of the dependence of the three key features - σ_a, μ_{42}^a, and μ_{42}^f - on SNR for only one realization of different modulation types of interest, based on the rule (2.22), for the ADMRA I are shown in Figs. 4.4 - 4.7. Also, sample results corresponding to this algorithm at the SNR of 15 dB and 20 dB are presented in this section. Furthermore, the overall success rates are introduced for the three algorithms.

A- Dependence of σ_a on SNR

σ_a is used to discriminate between the DSB and the PSK2 signal as well as between the combined (AM-FM) modulated and PSK4 signals. So, the dependence of σ_a on the SNR was computed for two digitally modulated signals - PSK2 and PSK4 and five analogue modulated signals - DSB, combined (Q=60%, D=5), combined (Q=60%, D=10), combined (Q=80%, D=5), and combined (Q=80%, D=10). The DSB and the combined modulated signals have amplitude information ($\sigma_a \geq t(\sigma_a)$), so it represent the subsets A. The PSK2 and the PSK4 signals have no amplitude variations in the non-weak intervals of a signal segment (constant instantaneous amplitude over the symbols durations) so they have no amplitude information ($\sigma_a < t(\sigma_a)$), and they represent the subset B.

For the discrimination between the DSB and PSK2 signals, the dependence of σ_a on the SNR is calculated for one realization of these two types as shown in Fig. 4.4. From Fig. 4.4, it is clear that for SNR > 0 dB the curve corresponding to PSK2 signal falls below the threshold level $t(\sigma_a)$= 0.25. Furthermore, for SNR > 0 dB the curve corresponding to the DSB is above the threshold $t(\sigma_a)$= 0.25.

For the discrimination between the combined and the PSK4 signals, the dependence of σ_a on the SNR is calculated for one realization of five modulated signals - PSK4, combined (Q=60%, D=5), combined (Q=60%, D=10), combined (Q=80%, D=5), and combined (Q=80%, D=10) - as presented in Fig. 4.5. From Fig. 4.5, it is clear that for SNR > 11 dB the curve corresponding to PSK4 signal falls below the threshold level $t(\sigma_a) = 0.15$. Furthermore, for SNR > 0 dB the curves corresponding to the other signals are above the threshold $t(\sigma_a)$= 0.15.

B- Dependence of μ_{42}^a on SNR

The dependence of μ_{42}^a on the SNR was computed for two digitally modulated signals - ASK2 and ASK4 - and two analogue modulated signals - AM (Q=60%) and AM (Q=80%). The AM signals have high "compactness instantaneous amplitude" ($\mu_{42}^a \geq t(\mu_{42}^a)$), so they represent the subset A while the ASK2 and the ASK4 signals have less "compactness instantaneous amplitude" ($\mu_{42}^a < t(\mu_{42}^a)$), so they represent the subset B.

The results for one simulation of these types are presented in Fig. 4.6. From Fig. 4.6, it is clear that for SNR ≥ 5 dB the curve corresponding to ASK2 signal falls below the threshold level $t(\mu_{42}^a) = 2.15$, and the curve corresponding to ASK4 falls bellow the threshold $t(\mu_{42}^a) = 2.15$ for SNR ≥ 10 dB. Furthermore, for SNR > 0 dB the curves corresponding to AM signals are above the threshold $t(\mu_{42}^a) = 2.15$.

C- Dependence of μ_{42}^f on SNR

The dependence of μ_{42}^f on the SNR was computed for two digitally modulated signals - FSK2 and FSK4 - and two analogue modulated signals - FM (D=5) and FM (D=10). The FM signals have high "compactness instantaneous amplitude" ($\mu_{42}^f \geq t(\mu_{42}^f)$), so they represent the subset A while the FSK2 and the FSK4 signals have less "compactness instantaneous amplitude" ($\mu_{42}^f < t(\mu_{42}^f)$), so they represent the subset B.

The results for one simulation of these types are presented in Fig. 4.7. From Fig. 4.7, it is clear that for SNR > 2 dB the curve corresponding to FSK2 signal falls below the threshold level $t(\mu_{42}^f) = 2.03$, and the curve corresponding to FSK4 falls below the threshold level $t(\mu_{42}^f) = 2.03$ for SNR > 2.5 dB. Furthermore, for SNR > 0 dB the curves corresponding to FM signals are above the threshold $t(\mu_{42}^f) = 2.03$.

4.4.2 Performance evaluations

The performance evaluations of the proposed ADMRAs are derived from 400 realizations for the twelve analogue modulated signals as well as the six digitally modulated ones. For the proposed ADMRA I, the performance results are summarised in Tables 4.3 - 4.6 for two values of SNRs (15 dB and 20 dB). In Tables 4.3 and 4.4, the ASK2 and the ASK4 signals are classified as one modulation type; MASK and also, the FSK2 and the FSK4 are classified as one modulation type; MFSK. Meanwhile the results in Table 4.5, represent the performance evaluation of the discrimination between the ASK2 and the ASK4 signals at the SNR of 15 dB and 20 dB. Also, the results in Table 4.6, represent the performance evaluation of the discrimination between the FSK2 and the

FSK4 signals at the SNR of 15 dB and 20 dB.

Extensive simulations of twelve analogue and six digitally modulated signals have been carried out at different SNR. Sample results have been presented at the SNR of 15 dB and 20 dB only. It was found that in the developed ADMRAs, using the decision-theoretic approach and at the SNR of 15 dB, all modulation types of interest have been classified with success rate $\geq 90.0\%$ except AM ($= 88.8\%$), ASK4 ($= 77.3\%$), and FSK4 ($= 88.0$ %). Finally, it is clear that the proposed algorithms introduce an improvement in the reduced SNR threshold over previous results reported in [5] and [16] as well as in the larger range of the signal modulation types considered.

4.4.3   Processing Time and Computational Complexity

Similar to the AMRAs and the DMRAs , the processing time and computational power are measured using the MATLAB software on the SUN SPARC station **20**. The values of the processing time (measured in seconds) and the computational power (measured in Megaflops) measured on the SUN SPARC station **20** are shown in Table 4.7. The processing time and the number of Megaflops, required to take a decision about the modulation type, correspond to only one signal segment in each case. The numbers in Table 4.7 are the average values of the measurements for 10 different realizations of each modulation type of interest at ∞ SNR.

4.5   Conclusions

The aim of the ADMRAs developed has been to distinguish automatically between twelve analogue and six digitally modulated signals. These algorithms use nine simple key features to perform the discrimination between this large number of modulation types, considered in this thesis, and they utilise the decision-theoretic approach. In the developed algorithms, the key feature threshold values are derived from 400 realizations of each modulated signal of interest at SNR of 15 dB and 20 dB. For the ADMRA I, the overall probability of correct decision at the SNR of 10 dB is about 88.7%.

Results for analogue and digital modulations recognition algorithms are summarised in Table 4.8 where each number represents the average probability of the correct classification of both the twelve analogue and the six digitally modulated signals at SNR of 15 dB and 20 dB. Using ADMRA I, the overall success rate for all the analogue and digitally modulated signals of interest is $\approx 95.0\%$ at the SNRs of 15 dB and 20 dB. Using ADMRAs II and III, it was found that the overall success rate for the recognition of all the analogue and digital modulations considered in this thesis is $\approx 93.0\%$ at the SNR of 15 dB and 20 dB. It is worth noting that, it is suggested that the threshold SNR for correct modulation recognition will reduce by applying the global procedure for modulation recognition (majority logic rule), explained in Section 2.2.

Modulation Type	Values of the Key Features at ∞ SNR		
	σ_a	μ_{42}^a	μ_{42}^f
AM	0.4	3.4	1.0
DSB	0.8	5.3	3.0
VSB	0.3	2.4	2.6
LSB	0.5	3.8	9.9
USB	0.5	3.8	9.9
Combined	0.4	3.4	2.8
FM	0.0	1.0	3.4
ASK2	0.5	1.5	1.0
ASK4	0.4	1.8	1.0
PSK2	0.1	2.2	3.6
PSK4	0.1	2.8	3.7
FSK2	0.0	1.0	1.4
FSK4	0.0	1.0	1.7

Table 4.1: Values of the key features σ_a, μ_{42}^a and μ_{42}^f for the ADMRAs *[based on one realization]*.

Key Features Thresholds	Optimum Value	$P_{av}(x_{opt})$	Notes
$t(\gamma_{max})$	[2-2.5]	100%	
$t(\sigma_{ap})$	$\pi/5.5$	100%	
$t(\sigma_{dp})$	$\pi/6$	99.9%	
$t(P)$	[0.6-0.9]	100%	SSB
	[0.5-0.7]	100%	VSB
$t(\sigma_a)$	[0.125-0.4]	100%	PSK2
	0.15	99.7%	PSK4
$t(\mu_{42}^a)$	2.15	87.3%	
$t(\mu_{42}^f)$	2.03	90.0%	
$t(\sigma_{aa})$	0.25	99.5%	
$t(\sigma_{af})$	0.40	99.8%	

Table 4.2: Optimum key features threshold values for the ADMRA I.

Simulated Types	Deduced Modulation Type										
	AM	DSB	VSB	LSB	USB	COM.	FM	MASK	PSK2	PSK4	MFSK
AM	88.8%	-	-	-	-	-	-	11.2%	-	-	-
DSB	-	100%	-	-	-	-	-	-	-	-	-
VSB	-	-	100%	-	-	-	-	-	-	-	-
LSB	-	-	-	100%	-	-	-	-	-	-	-
USB	-	-	1.7%	-	98.3%	-	-	-	-	-	-
COM.	-	-	-	-	-	100%	-	-	-	-	-
FM	-	-	-	-	-	-	90.0%	-	-	-	10.0%
ASK2	4.7%	-	-	-	-	-	-	95.3%	-	-	-
ASK4	22.7%	-	-	-	-	-	-	77.3%	-	-	-
PSK2	-	-	-	-	-	-	-	-	100%	-	-
PSK4	-	-	-	-	-	0.2%	-	-	-	99.8%	-
FSK2	-	-	-	-	-	-	8.0%	-	-	-	92.0%
FSK4	-	-	-	-	-	-	12.0%	-	-	-	88.0%

Table 4.3: Performance of the ADMRA I at SNR = 15 dB.

Simulated Types	Deduced Modulation Type										
	AM	DSB	VSB	LSB	USB	COM.	FM	MASK	PSK2	PSK4	MFSK
AM	86.1%	-	-	-	-	-	-	13.9%	-	-	-
DSB	-	100%	-	-	-	-	-	-	-	-	-
VSB	-	-	100%	-	-	-	-	-	-	-	-
LSB	-	-	0.5%	99.5%	-	-	-	-	-	-	-
USB	-	-	1.2%	-	98.8%	-	-	-	-	-	-
COM.	-	-	-	-	-	98.8%	-	-	-	1.2%	-
FM	-	-	-	-	-	-	90.0%	-	-	-	10.0%
ASK2	4.0%	-	-	-	-	-	-	96.0%	-	-	-
ASK4	19.8%	-	-	-	-	-	-	80.2%	-	-	-
PSK2	-	-	-	-	-	-	-	-	100%	-	-
PSK4	-	-	-	-	-	-	-	-	-	100%	-
FSK2	-	-	-	-	-	-	8.0%	-	-	-	92.0%
FSK4	-	-	-	-	-	-	12.0%	-	-	-	88.0%

Table 4.4: Performance of the ADMRA I at SNR = 20 dB.

Type	SNR = 15 dB		SNR = 20 dB	
	ASK2	ASK4	ASK2	ASK4
ASK2	98.3%	1.7%	100%	-
ASK4	0.2%	99.8%	-	100%

Table 4.5: Performance of the ADMRA I for discriminating ASK2 and ASK4.

Type	SNR = 15 dB		SNR = 20 dB	
	FSK2	FSK4	FSK2	FSK4
FSK2	99.5%	0.5%	100%	-
FSK4	0.5%	99.5%	-	100%

Table 4.6: Performance of the ADMRA I for discriminating FSK2 and FSK4.

Modulation Type	Computational power (Megaflops)	Processing time (Sec.)
AM	0.52	11.13
DSB	0.49	11.30
VSB	0.42	6.28
USB	0.42	6.11
LSB	0.42	6.07
Combined	0.58	11.29
FM	0.84	24.73
ASK2	0.55	11.40
ASK4	0.55	11.40
PSK2	0.50	11.96
PSK4	0.59	11.43
FSK2	1.02	35.80
FSK4	1.02	35.40

Table 4.7: Measured computational power and processing times of the ADMRA I on SPARC 20.

small

SNR (dB)	Overall performance	
	I	II & III
15	94.6%	93.3%
20	94.6%	93.2%

Table 4.8: Overall success rates for the developed ADMRAs.

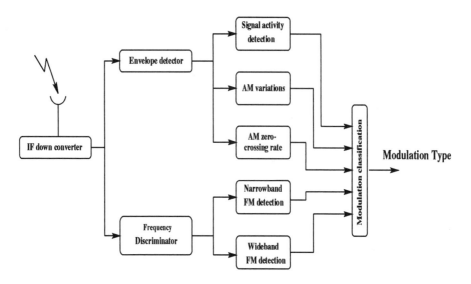

Figure 4.1: Simplified block schematic of *Petrovic* modulation recogniser [11].

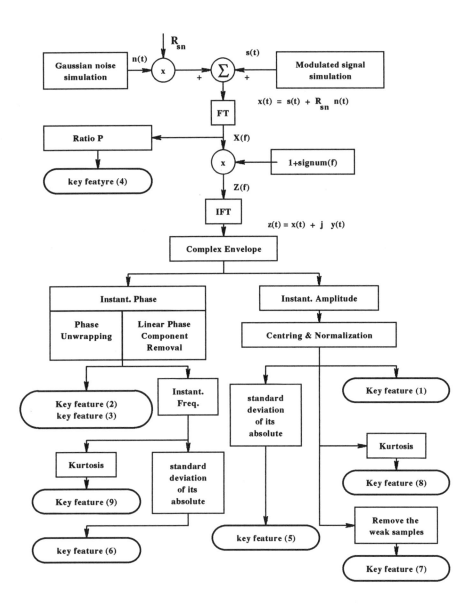

Figure 4.2: Functional flowchart for key features extraction for the ADMRAs.

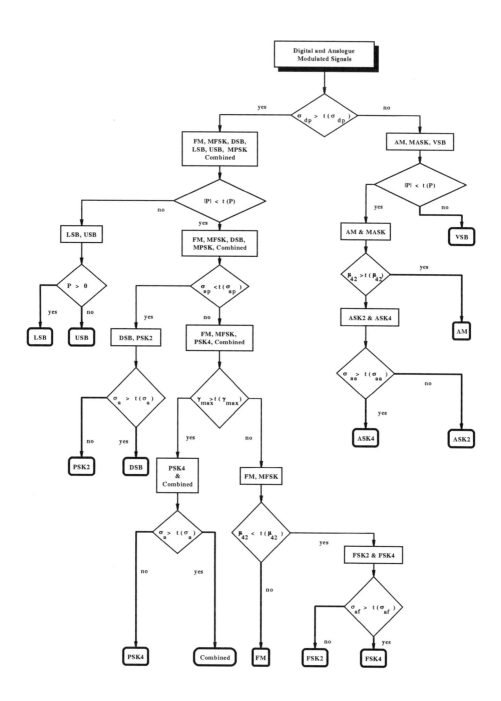

Figure 4.3: Functional flowchart for the ADMRA I.

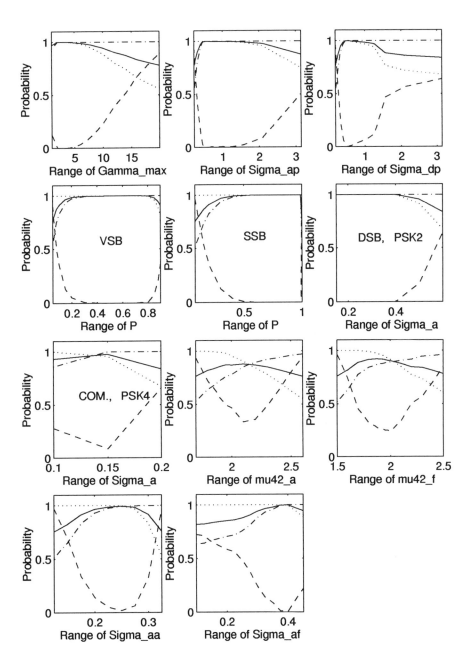

Figure 4.4: Key features thresholds determinations for the ADMRA I.

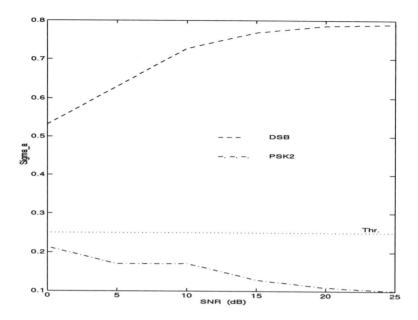

Figure 4.5: Dependence of σ_a (DSB & PSK2) on SNR for the ADMRA I.

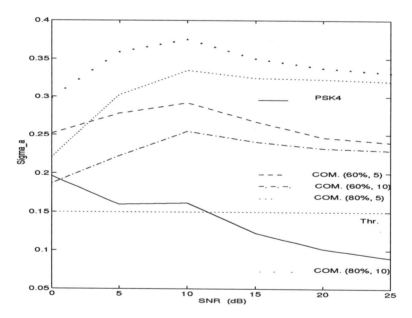

Figure 4.6: Dependence of σ_a (Combined & PSK4) on SNR for the ADMRA I.

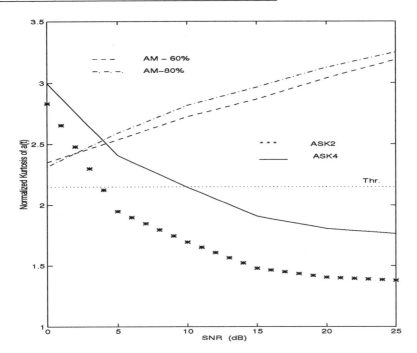

Figure 4.7: Dependence of μ_{42}^a on SNR for the ADMRA I.

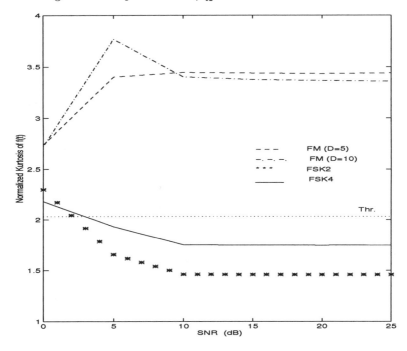

Figure 4.8: Dependence of μ_{42}^f on SNR for the ADMRA I.

Chapter 5

Modulation Recognition Using Artificial Neural Networks

5.1 Introduction

In this chapter the artificial neural networks (ANNs) approach as another solution for the modulation recognition process is studied in some detail. Unlike in other algorithms, especially those which utilise the decision-theoretic (DT) approach (Chapters 2 - 4), where a suitable threshold for each key feature has to be chosen, the threshold at each node (neuron) is chosen automatically and adaptively. Furthermore, in the DT approach, many algorithms based on the same key features can be developed by applying the extracted key features in different order in the classification algorithm and they perform with different success rates at the same SNR. In the DT algorithms, it was found that only one key feature is considered at a time. As a result, the probability of correct decision about a modulation type in these algorithms is based on the time-ordering of the key features used as well as probability of correct decision derived from each key feature. On the other hand, in the ANN algorithms all the key features are considered simultaneously. So, the time order of the key features does not affect on the probability of correct decision of on the modulation type of a signal. For that reason, it is suggested that the use of the ANN approach for solving the modulation recognition process may have better performance than the DT approach.

The fundamental principles of ANNs in the signal processing context have been previously described by many authors, e.g. Lippmann [39], Hush and Horne [40]. Furthermore, ANNs are successful in many practical applications including control [41], signal processing [42], pattern recognition [43], modelling [44], and manufacturing processes [45]. This work demonstrates the success of the ANN in the modulation recognition process. Due to the simplicity of the ANN structure, it can be used for on-line analysis. To use the ANNs for modulation recognition for on-line analysis, the main requirement is that the classifier structure be adjusted beforehand; i.e. to finish the training phase off-line before using this network in the on-line analysis.

In the next section introduces the suggested structure of the ANNs used in the modulation recognition process. In Section 5.3, the proposed algorithms for the recognition of analogue modulations only (AMRAs) are introduced. Section 5.4 presents the digital modulation recognition algorithms (DMRAs). Algorithms for the recognition of both analogue and digital modulations (ADMRAs) without any a priori information about the nature of a signal are introduced in Section 5.5. A summary of the results of the ANN algorithms in addition to comparisons with the decision-theoretic algorithms are introduced in Section 5.6. Finally, the chapter is concluded in Section 5.7.

5.2 Suggested Structure for ANN Modulation Recognisers

The neural network based modulation recognition process involves trying different architectures, learning techniques and training parameters in order to achieve an acceptable decision accuracy. The constraint that different inputs lead to the same output may lead to a network with hidden layers to learn certain mappings from input to output [40]. This is the situation in the modulation recognition process, especially for the analogue types such as the AM signals with different modulation depths that classify as AM, FM signals with different modulation indices that classify as FM, and combined (AM-FM) modulated signals with different modulation depths and different modulation indices that classify as combined modulation. The chosen ANN structure is different according to the group of modulation types under consideration - analogue only, digital only or both analogue and digital. The selection of the network parameters are based on choosing the structure that gives the minimum sum-squared errors (SSE) in the training phase, and the maximum probability of correct decisions in the test phase.

The modulation recognisers based on the ANN approach are composed of the three main blocks, as shown in Fig. 5.1. These are: 1) pre-processing in which the input key features are extracted from every signal realization in addition to signal isolation and segmentation, 2) training and learning phase to adjust the classifier structure, and 3) test phase to decide about the modulation type of a signal.

5.2.1 Pre-processing

Similar to the decision-theoretic algorithms, introduced in Chapters 2, 3, and 4, the intercepted signal of L seconds duration, sampled at the rate f_s, is divided into a set of adjacent segments each of length $N_s = 2048$ samples (equivalent to 1.707 msec.) and this results in $M_s = \frac{Lf_s}{N_s}$ segments. For every available segment the aforementioned key features, used in the corresponding DT algorithms, are measured. In the ANN algorithms introduced in this chapter, the input datasets are exactly the same as those used in the decision-theoretic algorithms. For example; the ANN analogue modulation recognition algorithms use the same four key features used in the decision-theoretic algorithms, presented in Chapter 2, and they are γ_{max}, σ_{ap}, σ_{dp}, and P. For digital modulation recognition using ANNs, the same five key features used in the decision-theoretic algorithms, introduced in Chapter 3, are used here in the ANN algorithms and they are γ_{max}, σ_{ap}, σ_{dp}, σ_{aa}, and σ_{af}. For both analogue and digital modulations recognition using ANNs, the same nine key features used in the decision-theoretic algorithms, presented in Chapter 4, are used here in the ANN algorithms and they are γ_{max}, σ_{ap}, σ_{dp}, P, σ_a, μ_{42}^a, μ_{42}^f, σ_{aa}, and σ_{af}. It is worth noting that the extraction of these key features is introduced in the appropriate places in Chapters 2, 3, and 4.

Furthermore, some of the M_s segments available are used in the training phase to adjust the ANN structure, while the others are used in the test phase to measure the performance and to decide about the modulation type. In the proposed ANN analogue modulation recognition and the digital modulation recognition algorithms from each simulated modulated signal (400 realizations each at 20 dB and at 10 dB), the first 50 realizations at 20 dB and 50 realizations at 10 dB are used in the training phase. In the recognition of both analogue and digital modulation algorithms using ANNs, from each simulated modulated signal (400 realizations each at 20 dB and at 15 dB), the first 50 realizations at 20 dB and 50 realizations at 15 dB are used in the training phase. Meanwhile, in the test phase, all the simulated realizations for each modulation type have been evaluated.

5.2.2 Training and learning phase of ANNs

The objective of training a network is to find the optimum weights and biases to minimise the error between the network output and the correct response. There are many types of learning methods to achieve the minimum error [46]. These are error-correction learning (back propagation), Hebbian learning, competitive learning, and Boltzman learning. A popular criterion is the minimum mean squared error between the network output and the correct response. Also, there are many learning paradigms such as the supervised, the unsupervised, and the self-organised learning [46]. In supervised learning [46], the training data must be provided in terms of input/output pairs denoted as $[\mathbf{X}, \mathbf{T}] = \{[\mathbf{x_1}, \mathbf{t_1}], [\mathbf{x_2}, \mathbf{t_2}], \ldots, [\mathbf{x_{Qs}}, \mathbf{t_{Qs}}]\}$, where $\mathbf{x_i}$ is a (Rix1) vector and Ri is the number of key features used (number of nodes in the input layer), $\mathbf{t_i}$ is a (Ox1) vector, O is the number of output decisions (number of nodes in the output layer), and Qs is the number of training pairs.

Both the back propagation as well as the supervised learning paradigm are used in all the developed ANN modulation recognition algorithms introduced in this chapter. The chosen ANNs are adaptively trained using momentum back propagation learning. In these algorithms, three network types are considered: 1) ANNs with no hidden layers, 2) ANNs with single hidden layer, and 3) ANNs with two hidden layers. All the networks used in the modulation recognition algorithms in this chapter are adaptively trained to reduce the sum squared error, SSE, defined in terms of the difference between the calculated output corresponding to the data used in the training and the actual target (defined for each group of algorithms - AMRAs, DMRAs, and ADMRAs - in Sections 5.3.2, 5.4.2 and 5.5.2). The training SSE for the three network types is defined as follows

1. In the networks with no hidden layers (input and output layers only), the SSE is defined by

$$SSE = \sum_{i=1}^{O} \sum_{j=1}^{Qs} E(i,j), \tag{5.1}$$

where

$$\mathbf{E} = (\mathbf{T} - \mathbf{A})^2 \tag{5.2}$$

$\mathbf{T}$ is the actual target and $\mathbf{A}$ is the calculated target and is given by

$$\mathbf{A} = \text{log-sigmoid}\,(\mathbf{W}\mathrm{x}\mathbf{P_{in}} + \mathbf{B}) \tag{5.3}$$

where, $\mathbf{W}$ and $\mathbf{B}$ are the weights and biases of the output layer, and $\mathbf{P_{in}}$ is the input data used for training. The log-sigmoid function is defined as follows

$$\begin{aligned}
\text{log-sigmoid}\,(\mathbf{Y}) &= [\text{log-sigmoid}\,(y_{ij})]_{\text{mxn}} \\
&= \left[\frac{1}{1 + e^{-y_{ij}}}\right]_{\text{mxn}} \tag{5.4}
\end{aligned}$$

where $\mathbf{Y}$ is a (mxn) matrix.

The dimensions of all the matrices and vectors used can be expressed as follows: $\mathbf{P_{in}}$, $\mathbf{W}$, $\mathbf{B}$, $\mathbf{b}$, $\mathbf{A}$, are respectively (RixQs), (OxRi), (OxQs), (Ox1), and (OxQs) matrices and $\mathbf{B}$ is given by $[\mathbf{b}, \ldots, \mathbf{b}]$.

2. In the networks with one hidden layer, SSE is as defined by (5.1), but (5.2) is re-expressed as

$$\mathbf{E} = (\mathbf{T} - \mathbf{A_2})^2 \tag{5.5}$$

where $\mathbf{T}$ is the actual target and $\mathbf{A_2}$ is the calculated target, and is given by

$$\mathbf{A_2} = \mathbf{W_2}\mathrm{x}\mathbf{A_1} + \mathbf{B_2} \tag{5.6}$$

where $\mathbf{A_2}$ is the output of the second (output) layer containing O nodes, $\mathbf{W_2}$ and $\mathbf{B_2}$ are the weights and biases of the second layer, $\mathbf{A_1}$ is the output of the first (hidden) layer and can be expressed as

$$\mathbf{A_1} = \text{log-sigmoid}(\mathbf{W_1}\mathrm{x}\mathbf{P_{in}} + \mathbf{B_1}), \tag{5.7}$$

where $\mathbf{W_1}$ and $\mathbf{B_1}$ are the weights and biases of the first (hidden) layer containing S nodes. The activation function associated with the ouput layer in this network is a linear function, which is defined by

$$\text{linear}(\mathbf{Y}) = [\text{linear}(y_{ij})]_{\text{mxn}}$$
$$= [y_{ij}]_{\text{mxn}}, \tag{5.8}$$

The dimensions of all the matrices and vectors used can be expressed as follows: $\mathbf{P_{in}}$, $\mathbf{W_1}$, $\mathbf{B_1}$, $\mathbf{b_1}$, $\mathbf{A_1}$, $\mathbf{W_2}$, $\mathbf{B_2}$, $\mathbf{b_2}$, and $\mathbf{A_2}$, are respectively (RixQs), (SxRi), (SxQs), (Sx1), (SxQs), (OxS), (OxQs), (Ox1), and (OxQs) matrices and $\mathbf{B_1}$ is given by $[\mathbf{b_1}, \ldots, \mathbf{b_1}]$ while $\mathbf{B_2}$ is given by $[\mathbf{b_2}, \ldots, \mathbf{b_2}]$.

3. In the networks with two hidden layers, the SSE is as given by (5.1), but (5.2) is re-expressed as

$$\mathbf{E} = (\mathbf{T} - \mathbf{A_3})^2 \tag{5.9}$$

where $\mathbf{T}$ is the actual target and $\mathbf{A_3}$ is the calculated target (output of the third layer) and is given by

$$\mathbf{A_3} = \text{log-sigmoid}(\mathbf{W_3}\text{x}\mathbf{A_2} + \mathbf{B_3}) \tag{5.10}$$

where $\mathbf{W_3}$ and $\mathbf{B_3}$ are the weights and biases of the third (output) layer, $\mathbf{A_2}$ is the output of the second hidden layer and can be expressed as

$$\mathbf{A_2} = \mathbf{W_2}\text{x}\mathbf{A_1} + \mathbf{B_2} \tag{5.11}$$

where $\mathbf{W_2}$ and $\mathbf{B_2}$ are the weights and biases of the second hidden layer, $\mathbf{A_1}$ is the output of the first hidden layer and can be expressed as

$$\mathbf{A_1} = \text{log-sigmoid}(\mathbf{W_1}\text{x}\mathbf{P_{in}} + \mathbf{B_1}) \tag{5.12}$$

where $\mathbf{P_{in}}$ is the input data used for training, and $\mathbf{W_1}$ and $\mathbf{B_1}$ are the weights and biases of the first hidden layer.

Let the number of nodes in the first hidden-layer be S1, the number of nodes in the second hidden-layer be S2, and the number of realizations used for training be Qs. The dimensions of all the matrices and vectors used can be expressed as follows: $\mathbf{P_{in}}$, $\mathbf{W_1}$, $\mathbf{B_1}$, $\mathbf{b_1}$, $\mathbf{A_1}$, $\mathbf{W_2}$, $\mathbf{B_2}$, $\mathbf{b_2}$, $\mathbf{A_2}$, $\mathbf{W_3}$, $\mathbf{B_3}$, $\mathbf{b_3}$, and $\mathbf{A_3}$, are respectively (RixQs), (S1xRi), (S1xQs), (S1x1), (S1xQs), (S2xS1), (S2xQs), (S2x1), (S2xQs), (OxS2), (OxQs), (Ox1), and (OxQs) matrices, and $\mathbf{B_1}$ is given by $[\mathbf{b_1}, \ldots, \mathbf{b_1}]$, $\mathbf{B_2}$ is given by $[\mathbf{b_2}, \ldots, \mathbf{b_2}]$ while $\mathbf{B_3}$ is given by $[\mathbf{b_3}, \ldots, \mathbf{b_3}]$.

The output of the training phase are the weights and biases of the output layer of the trained network that will be used in the test phase. The training phase of the networks with single and two hidden layers are depicted in Fig. 5.2. Generally the training phase comprises the following steps:

1. **Initialise the network weights and biases**: The numbers of hidden layers as well as the number of nodes in each hidden layer along with the training parameters such as the maximum epochs, the required error goal, the learning rate, the error ratio, the learning rate increment and the learning rate decrement are chosen. The initial weights - $\mathbf{W}_1^0$ and $\mathbf{W}_2^0$ (single hidden layer) or $\mathbf{W}_1^0$, $\mathbf{W}_2^0$, and $\mathbf{W}_3^0$ (two hidden layers) - and similarly the initial biases - $\mathbf{B}_1^0$ and $\mathbf{B}_2^0$ (single hidden layer) or $\mathbf{B}_1^0$, $\mathbf{B}_2^0$, and $\mathbf{B}_3^0$ (two hidden layers) - of the network are chosen (see Fig. 5.2) using the random number generator.

2. **Adaptive training stage**: To decrease the training time an adaptive learning is used. The adaptive learning is characterised by three parameters. These are: the learning rate (lr), the rate increment (lr_{inc}) and the rate decrement (lr_{dec}). In the developed algorithm, these parameters are chosen to be 10^{-5}, 1.05 and 0.7 respectively. This stage consists of the following steps:

   (a) **Define the network parameters**: in which the suitable activation functions are chosen. In the proposed ANN algorithms, the log-sigmoid and the linear functions are used as two activation functions for the different ANN types - no hidden layer, one hidden layer, and two hidden layers ANNs. The activation functions associated with the different layers in the three network types are used as follows:

   - In the no hidden layer ANNs, the log-sigmoid function is the activation function associated with the output layer.

   - In the single hidden-layer ANN algorithms, the activation function associated with the hidden layer is the log-sigmoid function, and the activation function associated with the output layer is the linear function.

   - In the two hidden layers ANN algorithms, the activation function associated with the first hidden layer is the log-sigmoid function, the activation function associated with the second hidden layer is the linear function, and the activation function associated with the output layer is the log-sigmoid function.

(b) **Presentation phase**: in which the sum squared error, SSE, of the difference between the calculated output corresponding to the data used in the training and the actual output is computed as given (5.1), and using (5.2), (5.5) and (5.9) to define **E**.

(c) **Initialise the changes in weights and biases of the network.**

(d) **Back propagation with momentum and adaptive learning**:
The momentum allows the network to respond not only to the local minimum but also to the recent trends in the error surface. Without the momentum, a network may get stuck in a shallow local minimum but with the momentum a network can slide through such a minimum. Typically the momentum constant, m_c, is set to 0.95.

Also, in the adaptive learning if the present error exceeds the previous error by more than a pre-defined ratio (typically $= 1.04$), the new weights, biases, outputs and error are discarded and the learning rate is decreased (multiply by 0.7). If the present error is less than the previous error, the new values are kept and the learning rate is increased (multiply by 1.05).

3. **Presentation phase**: The output of this stage is based on the ANN type and it defined as follows:

- For no hidden layer ANNs, the output of this stage is **A**, and the SSE. **A** is calculated as expressed in (5.3).

- For one hidden layer ANNs, the output of this stage as shown in Fig. 5.2a is **A₁**, **A₂**, and the SSE. **A₂** and **A₁** are calculated as expressed in (5.6) and (5.7) respectively.

- For two hidden layers ANNs, the output of this stage as shown in Fig. 5.2b is **A₁**, **A₂**, **A₃** and the SSE. **A₃**, **A₂** and **A₁** are calculated as expressed in (5.10), (5.11) and (5.12) respectively.

4. **Back propagation learning with momentum only**: For more reduction of the SSE of the network, momentum learning is used. The following steps in every epoch are executed

- **Check phase**: For every epoch, the calculated SSE is compared with the pre-chosen error goal as

  If $SSE \leq$ the error goal, the learning stops. In this case the weights and biases of the network are saved for future use in the test phase. On the other hand, if $SSE >$ the error goal the algorithm is continued till the maximum number of epochs is completed.

- **Updating the network weights and biases**: In the main part of the momentum learning stage, the updating of the network weights is based on the following rule:

$$\mathbf{W_n^{(i)}} = \mathbf{W_n^{(i-1)}} + \mathbf{\Delta W_n^{(i)}} \tag{5.13}$$

  where $n = 1$, 2 or 3, i=1,2,..., K, K is the maximum number of epochs, $\mathbf{\Delta W_n}$ are the updates in the weights.

- **Presentation phase**: It is exactly the same as explained above.

5.2.3 Test phase of ANNs

In the ANN test phase, the only data needed from the trained networks are the synaptic weights and biases which must be available for the use in the test phase of the trained ANN. In the AMRAs and the DMRAs, the training has been done using only 50 realizations at each SNR (10 dB and 20 dB) for each of the twelve analogue and the six digitally modulated signals (presented in Chapters 2, 3, and 4), based on the kind of algorithms - AMRAs or DMRAs - used, while the performance is measured for the 400 realizations of each modulation type of interest and at two SNR values (10 dB and 20 dB). In the ADMRAs, the training has been done using only 50 realizations at each SNR (15 dB and 20 dB) for each analogue and digitally modulated signal of interest, while the performance is measured for the 400 realizations of each modulation type of interest and at two SNR values (15 dB and 20 dB). Generally, the test phase or the performance evaluations as shown in Fig. 5.3 is composed of the following steps:

1. the key features of the set of realizations to be used in the test phase are introduced to the trained network.

2. the actual target matrix, **T**, is defined for each group of modulations as will be explained later in this chapter.

3. for every realization of the test group, the corresponding output vector, based on the number of hidden layers used, is computed as follows.

4. the element corresponding to the maximum value in the output vector is set to **1** and the other elements are set to **0**.

5. the modified output vector should correspond to one of the columns of the matrix **T** and this correspondence is taken as the deduced modulation type.

6. repeat the whole procedure for each realization in the test group.

7. the probability of correct decision is computed as the percentages of realizations that have the correct modified output vector.

The performance evaluations of the proposed algorithms, utilising the ANN approach will be introduced for the same twelve analogue modulated signals - AM ($Q = 60\%$), AM ($Q = 80\%$), DSB, VSB, LSB, USB, combined ($Q = 60\%$, $D = 5$), combined ($Q = 60\%$, $D = 10$), combined ($Q = 80\%$, $D = 5$), combined ($Q = 80\%$, $D = 10$), FM ($D=5$), FM ($D=10$) - presented in Chapter 2, and the six digitally modulated signals - ASK2, ASK4, PSK2, PSK4, FSK2, FSK4 - presented in Chapters 3, for direct comparisons between the decision-theoretic and the ANN algorithms. Three categories of the modulation recognisers are presented in the rest of this chapter. The first is for the recognition of analogue modulations only, the second is for the recognition of digital modulations only, and the third is for the recognition of both analogue and digital modulations. In the first and the second categories, two ANN types are considered - one hidden layer and two hidden layers. The performance evaluations of the developed algorithms for the recognition of either analogue modulations only or digital modulations only are measured at the SNR of 10 dB and 20 dB. In the third category, the performance of the developed algorithm for both analogue and digital modulations is evaluated at the SNR of 15 dB and 20 dB.

5.3   Analogue Modulation Recognition Algorithms (AMRAs)

The developed algorithms for analogue modulations recognition is based on using single and double hidden layer ANNs. The difference between these two algorithms is in the number of hidden layers used. It is found that the two hidden layer ANNs offers better performance than the single hidden layer. Furthermore, the reduction of the training and learning time for these algorithms will be considered in this section. The modulation types that can be classified by these recognisers are: AM, DSB, VSB, LSB, USB, combined, and FM signals.

5.3.1   Choice of ANN architectures

Two network types, based on the number of hidden layers, are considered for the AMRAs. These are: 1) single hidden layer algorithms and 2) two hidden layers algorithms. A lot of work has been done to choose the optimum network structure. All the tested networks contain a 4-node input layer, a 7-node output layer, and they differ in the number of hidden layers and the number of nodes in each hidden layer. Through the rest of this chapter, any double hidden layer ANN used for modulation recognition is denoted by (S1-S2). This means that this network has S1 nodes in the first hidden layer, and S2 nodes in the second hidden layer.

In the single hidden layer AMRAs [47], all the tested networks contain a 4-node input layer, a 7-node output layer, and they are differ in the number of the nodes in the hidden layer. The numbers in Table 5.1 are based on 400 simulations of each of the twelve analogue modulations at SNRs of 10 dB as well as 20 dB. Table 5.1 shows that the 25-node hidden layer obtains the largest probability of correct decisions on the analogue modulation types than the other networks (20- and 30-node). Dependence of the SSE on the number of epochs for different number of nodes in the hidden layer for the AMRAs is displayed in Fig. 5.4. It is clear that a 25-node hidden layer for the analogue modulation recognition is better than choosing a 20- or a 30-node hidden layer network with respect to the SSE. So, a network with a 4-node input layer, a 25-node hidden layer, and a 7-node output layer as shown in Fig. 5.5 is considered further to evaluate the performance and to reduce the training time of the single hidden layer ANN algorithm for the analogue modulations recognition.

In the double hidden layer analogue modulation recognition algorithms [48], the choice of the ANN architectures is made to give higher performance and less SSE than those of the single hidden layer ANNs. The numbers in Table 5.2 are based on 400 simulations of each analogue modulated signal of interest at the SNR of 10 dB, and 20 dB for 100,000 training epochs only. From Table 5.2 it is observed that six of the tested networks perform well and they achieve overall success rate > 99.0%. The numbers in Table 5.3 are based on the same 400 simulations of each of the twelve analogue modulated signals at the SNR of 10 dB as well as 20 dB and for 100,000 epochs for both training and learning. It is clear that five of the best six networks still achieve success rate > 99.0% but the overall success rate for one, (7-5), of them is reduced (=97.6%) after 100,000 learning epochs. The dependence of the learning SSE on the number of epochs for the best five network structures, that achieve success rate >99.0%, is displayed in Fig. 5.6. The best five networks are: (5-7), (7-7), (10-7), (10-10), and (12-12). The selection of the network parameter are based on choosing the structure that gives the minimum SSE and the maximum probability of correct decisions. In this case the two hidden layers network with 10 nodes in each of the hidden layers as shown in Fig. 5.7 is considered further to evaluate the performance and to reduce the training time of the double layer ANN algorithm for analogue modulation recognition.

5.3.2 Performance evaluations

In the developed AMRAs utilising the ANN approach, the actual target, **T**, is the (7x7) identity matrix. In this matrix, the columns in the ascending order correspond to the AM decision, the DSB decision, the VSB decision, the LSB decision, the USB decision, the combined decision, and the FM decision respectively. In this section, the performance of two ANN algorithms - single and double hidden layers - for analogue modulations recognition are evaluated.

For the single hidden layer ANN analogue modulation recognition algorithm, the results of the performance evaluation of the proposed procedure for analogue modulation recognition, using the single hidden layer (25-node) ANN and derived from 400 realizations for each analogue modulated signal of interest, are summarised in Tables 5.4 and 5.5 for two values of SNRs (10 dB and 20 dB). From Table 5.4, it is clear that all the types of analogue modulations have been correctly classified with more than 95.0% success rate and three of them have been classified with success rate 100%. Excluding the USB, the success rate is > 98.0% at 10 dB for all analogue modulations. At the SNR of 20 dB (see Table 5.5) all the modulation types have been classified with success rate > 94.0%. Excluding the USB, the success rate > 99.0% and now five modulation types have been classified with success rate 100%.

For the two-hidden layers ANN analogue modulation recognition algorithm, the results of the performance evaluation of the proposed procedure for analogue modulation recognition, using the two hidden layers (10-10) ANN and derived from 400 realizations for each type of modulations, are summarised in Tables 5.6 and 5.7 for two values of SNRs (10 dB and 20 dB). From Table 5.6, it is clear that all types of analogue modulation have been correctly classified with more than 98.0% success rate and two of them have been classified with success rate 100%. In Table 5.7 (SNR = 20 dB) all the modulation types have been classified with success rate > 99.0%, and now four modulation types have been classified with success rate 100%.

5.3.3   Speed-up of the training phase

The main requirement to use the ANN algorithms for on-line analysis is to adjust the classifier structure beforehand; i.e. to finish the training phase in the off-line analysis. Furthermore, it is found that all the tested networks required a long training time, as the key features values for different modulation types have different ranges. For example; the value of γ_{max} is of order 10^{-1} for FM signals and 10^2 for DSB signal. So, normalisation of the datasets is used to speed up the training and learning phase. Normalising the datasets reduces the range of their values to be $\leq +1$ and ≥ -1, and this leads to avoiding the problem of long training time. However, the normalisation should be applied in the test phase as well as in the training phase.

In the training phase, the data corresponding to each key feature are normalised with respect to the maximum value over all the segments used in the training phase. In the test phase, two cases are considered: 1) all the segments, that used in the test phase, are available at the beginning (off-line analysis), and 2) not all the segments are available at the beginning, only one segment is available at a time, (on-line analysis). Thus, three suggested ways for the normalisation in the test phase are considered. These are: 1) the maximum value is taken over all the test data (off-line analysis only), and 2) the maximum value of the training data is consider as the maximum for the test data, 3) the maximum value of the test data is consider as the maximum value of the training data then updating this maximum value by comparing the value of the key feature for every segment with the previous maximum value considered. Note that the initial maximum value is taken as the maximum value over the datasets used for training the network.

Fig. 5.8 represents the SSE values for only 100,000 training epochs for the single hidden layer (25 nodes) ANN algorithm for analogue modulations recognition, with and without normalisation. From Fig. 5.4 and Fig. 5.8, it is clear that applying the normalisation introduces SSE at 100,000 training epochs less than the SSE for the same network structure (25 nodes) in which the normalisation is not used, and that derived from 100,000 epochs for both training and learning; the speed increases twice at least. So, normalising the key features reduces the training time in the AMRA with a single hidden layer.

Fig. 5.9 represents the SSE values for only 50,000 training epochs for the two hidden layers (10-10) ANN algorithm for analogue modulation recognition, with and without normalisation. From Fig. 5.6 and Fig. 5.9, it is clear that applying the normalisation introduces SSE at 50,000 training epochs less than the SSE for the same network structure (10-10) in which the normalisation is not used, and that derived from 100,000 epochs for both training and learning; the speed increases 4 times at least. So, normalising the key features reduces the training time in the AMRA with two hidden layers.

It worth noting that the overall success rates are measured for the single hidden layer (25 nodes) and the double hidden layer (10-10) ANN algorithms. In the case of a single hidden layer and without normalisation, choosing the network with 25 nodes in its hidden layer as the best over all the tested networks is done after 100,000 epochs for both training and learning. In the case of the two hidden layers and without normalisation, choosing the network (10-10) as the best over all the tested networks is done after 100,000 epochs for both training and learning. After applying the normalisation and in the case of a single hidden layer ANN algorithm, only 100,000 training epochs is adequate to achieve the minimum SSE or less for that obtained with no normalisation. Furthermore, the overall success rate of the single hidden layer (25 nodes) ANN algorithm for analogue modulations recognition and in which the normalisation is applied is $\approx$ 96.0% at the SNRs of 10 dB and 20 dB SNR. Meanwhile, in the case of the two hidden layers ANN algorithm, only 50,000 training epochs is sufficient to achieve the same minimum SSE or less for that obtained in the no normalisation case. Furthermore, the overall success rate of the two hidden layers (10-10) ANN algorithm for analogue modulations recognition and in which the normalisation is applied is $\approx$ 98.0% at the SNR of 10 dB and at 20 dB SNR.

5.4 Digital Modulation Recognition Algorithms (DM-RAs)

This section is concerned with presenting two ANN algorithms for digital modulation recognition. The difference between these two algorithms is in the number of hidden layers used. It is found that the two hidden layers ANN offers better performance than the single hidden layer. Also, the reduction of the training time for these algorithms is considered in this section. The modulation types which can be classified by these recognisers are: ASK2, ASK4, PSK2, PSK4, FSK2, and FSK4 signals.

5.4.1   Choice of ANN architectures

Similar to the AMRAs using the ANNs, two network types, based on the number of the hidden layers, are considered for the recognition of digitally modulated signals, and they are the single hidden layer networks and the double hidden layer networks. Also, much work has been done in the two cases - single and double hidden layers - to choose the optimum network over all those tested.

In the single hidden layer DMRAs [47, all the tested networks contain a 5-node input layer, a 6-node output layer, and they are differ in the number of nodes in the hidden layer. The numbers in Table 5.8 are based on 400 simulations of each of the six digital modulations at SNR of 10 dB as well as 20 dB. Dependence of the SSE on the number of epochs for different numbers of nodes in the hidden layer for digital modulation recognition is displayed in Fig. 5.10. It is clear that a 10-node hidden layer for the digital modulation recognition is better than choosing a 5- or a 15-node hidden layer network with respect to the SSE. Also as shown in Table 5.8, it is clear that the 10-node hidden layer network gives higher probability of correct decision on the digital modulation types than the other networks (5- and 15-node). So, a network with a 5-node input layer, a 10-node hidden layer, and a 6-node output layer as shown in Fig. 5.11 is considered further to evaluate the performance and to reduce the training time of the single hidden layer ANN algorithm for digital modulation recognition.

In the double hidden layer digital modulation recognition algorithms [48], all the tested networks contain a 5-node input layer, a 6-node output layer, and they differ in the number of nodes in the two hidden layers. Similar to the two hidden layers AMRAs using ANNs, any two hidden layers ANN used for digital modulations recognition has a 5-node input layer, an S1-node first hidden layer, an S2-node second hidden layer, and a 6-node output layer. The numbers in Table 5.9 are based on 400 simulations for each of the digital modulation types of interest at the SNR of 10 dB, and 20 dB for 150,000 epochs training only. From Table 5.9 it is observed that five of the tested networks perform well and they achieve overall success rate $> 96.0\%$. The numbers in Table 5.10 are based on the same 400 simulations of each of the six digital modulations at the SNR of 10 dB as well as 20 dB and for 150,000 epochs for both training and learning for the best five networks. The dependence of the learning SSE on the number of epochs for these best networks structures is displayed in Fig. 5.12. The best five networks are: (7-5), (7-7), (10,7), (10-10), and (12-12). The selection of the network parameter are based on choosing the structure that gives the minimum SSE and the maximum probability of correct decisions. In this section the two hidden layers network with 12 nodes in each of the hidden layers as shown in Fig. 5.13 is considered further to evaluate the performance of the two hidden layers ANN algorithm for digital modulation recognition.

5.4.2   Performance evaluations

In the developed DMRAs, the actual target, **T**, is the (6x6) identity matrix. In this case, the columns in the ascending order correspond to the ASK2 decision, the ASK4 decision, the PSK2 decision, the PSK4 decision, the FSK2 decision, and the FSK4 decision respectively.

For the single hidden layer ANN algorithm for digital modulation recognition, the results of the performance evaluation of the proposed procedure for digital modulation recognition, using the ANN and derived from 400 realizations for each type of modulations, are summarised in Tables 5.11 and 5.12 for two values of SNR (10 dB and 20 dB). From Table 5.11, it is clear that all types of digital modulations have been correctly classified with more than 93.0% success rate. Excluding the ASK2, all the other types have been classified with success rate > 95.0%. At the SNR of 20 dB (see Table 5.12) all the modulation types have been classified with success rate > 97.0%, and now four modulation types have been correctly classified all the time (100% success rate). Furthermore, the overall success rate is ≈ 97.0% at the SNR of 10 dB and it is = 99.6% at 20 dB SNR.

For the double hidden layer ANN algorithms, the results of the performance evaluation of the proposed procedure for digital modulation recognition are summarised in Tables 5.13 and 5.14 for two values of SNRs (10 dB and 20 dB). From Table 5.13, it is clear that all the digital modulation types have been correctly classified with more than 96.0% success rate except FSK2 (= 92.5%). At the SNR of 20 dB as shown in Table 5.14, all the modulation types have been classified with success rate > 99.0%, and four of them have been correctly classified all the time (100% success rate).

5.4.3   Speed-up of the training phase

Applying the normalisation as presented in the AMRAs, a reduction in the training time can be observed. Fig. 5.14 represents the SSE values for only 100,000 training epochs for the single hidden layer (10 nodes) ANN algorithm for digital modulations recognition, with and without normalisation. From Fig. 5.10 and Fig. 5.14, it is clear that applying the normalisation introduces SSE at 100,000 training epochs less than the SSE for the same network structure (10 nodes) in which the normalisation is not used, and that derived from 150,000 epochs for both training and learning; the speed increases at least 3 times. So, normalising the key features reduces the training time in the DMRA with single hidden layer ANN.

Fig. 5.15 represents the SSE values for only 50,000 training epochs for the double hidden layer, 12 nodes in each hidden layer, with and without normalisation. From Fig. 5.12 and Fig. 5.15, it is clear that applying the normalisation introduces SSE at 50,000 training epochs less than the SSE for the same network structure (12-12) without normalisation and using 150,000 epochs for both training and learning. So, the normalisation increases the speed of the training phase at least 6 times with respect to that without normalisation.

Also, the overall success rates are measured for the single hidden layer (10 nodes) and the double hidden layer (12-12) ANN algorithms. In the case of a single hidden layer and without normalisation, choosing the network with 10 nodes in its hidden layer as the best over all the tested networks is done after 150,000 epochs for both training and learning. In the case of two hidden layers and without normalisation, choosing the network (12-12) as the best over all the tested networks is done after 150,000 epochs for both training and learning. After applying the normalisation and in the case of a single hidden layer ANN algorithm, only 100,000 training epochs is adequate to achieve the minimum SSE or less for that obtained with no normalisation. Furthermore, the overall success rate of the single hidden layer (10 nodes) ANN algorithm for digital modulation recognition and in which the normalisation is applied, is $\approx 96.0\%$ at the SNR of 10 dB and it is $\approx 99.0\%$ at 20 dB SNR. Meanwhile, in the case of the two hidden layers ANN algorithm, only 50,000 training epochs is sufficient to achieve the same minimum SSE or less for that obtained from the no normalisation case. Furthermore, the overall success rate of the two hidden layers (12-12) ANN algorithm for analogue modulations recognition and in which the normalisation is applied, is $\approx 97.0\%$ at the SNR of 10 dB and it is $\approx 100\%$ at 20 dB SNR.

5.5   Analogue and Digital Modulations Recognition Algorithms (ADMRAs)

The developed ADMRA is based on using a three network structure [48]. The first is a two hidden layer ANN, and the second and the third are very simple and comprise only a one node input layer, a two nodes output layer and no hidden layers. Furthermore, key features normalisation is considered to speed up the training phase.

5.5.1   Choice of the ANN architectures

It was found that the best ANN structure, as shown in Fig. 5.16, contains three networks. Only two of the output decisions of the first network require an additional two networks to complete the discrimination. These two decisions are the MASK decision and the MFSK decision. The second and the third networks are used to estimate the number of levels, M, of the MASK and MFSK signals. For the first network, it was found that the best ANN possesses three neuron layers. The second and the third networks are very simple as they just comprise a one node input layer and a two node output layer network (no hidden layers).

First network structure

A lot of work has been done to choose the optimum structure for the first network, i.e. the number of nodes in the hidden layer, and the number of epochs in training and learning phases. The appropriate network architecture has been chosen such that the number of the input layer nodes equals to the number of the extracted key features and the number of the output layer nodes equals the number of desired decisions. In the ANN algorithm developed for both analogue and digital modulations recognition, many structures can perform well out of all the tested structures for the first network. All of them contain a 7-node input layer, a 11-node output layer, and two hidden layers. The only difference between these structures is in the number of nodes in the hidden layers. It is worth noting that for the first network, any network structure used is denoted by (S1-S2). This means that this network has a 7-node input layer, S1 nodes in the first hidden layer, S2 nodes in the second hidden layer, and an 11-node output layer.

The selection of the network parameters is based on choosing the structure that gives the minimum SSE and the maximum probability of correct decisions. The numbers in Table 5.15 are based on 400 simulations of each of the twelve analogue and the six digitally modulated signals at SNR of 15 dB as well as 20 dB and for 100,000 training epochs. Table 5.16 shows the overall performance of the best six networks (success rate > 93.0%) based on 250,000 epochs for both the training and the learning. The dependence of the learning SSE on the number of epochs for these six networks structures is displayed in Fig. 5.17. It is clear that four of these six structures give overall performance > 95.0% as shown in Table 5.16. Thus, the best four networks are (12-12), (12-15), (15-15), and (18-18). The network with 15 nodes in each of the hidden layers as shown in Fig. 5.18 is considered further to evaluate the performance and to reduce the training time of the first network of the proposed ADMRA, using the ANN approach.

Second network structure

This network is used when the output decision of the first network is MASK. It is found that the second network is very simple network and it comprises a one node input layer and a two node output layer (no hidden-layers). The input of this network is σ_{aa} and its output is the ASK2 or the ASK4 decision.

Third network structure

This network is used when the output decision of the first network is MFSK. It is found that this network has the same structure as the second network. The input of this network is σ_{af} and its output is the FSK2 or the FSK4 decision.

5.5.2 Performance evaluations

In the developed ADMRA, as it is shown in Fig. 5.16, the best choice for the ANN algorithm for both analogue and digital modulations recognition, comprises three networks, So the actual target is defined for each of them as follows

- For the first ANN (two hidden layers), **T** is the (11x11) identity matrix. In this matrix, the columns in the ascending order correspond to the AM decision, the DSB decision, the VSB decision, the LSB decision, the USB decision, the combined decision, the FM decision, the PSK2 decision, the PSK4 decision, the MASK decision, and MFSK decision respectively.

- For the second and the third ANNs (no hidden layers), **T** is the (2x2) identity matrix. These networks are used to discriminate either between the ASK2 and the ASK4 as well as between the FSK2 and the FSK4 modulations. In first ANN, if the decision is MASK (or MFSK) a very simple network is used with the first column of **T** corresponding to the ASK2 (or FSK2) decision, and the second column corresponding to the ASK4 (or FSK4) decision.

The performance evaluations of the proposed ADMRA (15-15) are introduced for twelve analogue modulated signals as well as six digitally modulated ones. The results of the performance evaluations are summarised in Tables 5.17 - 5.20. In Tables 5.17 and 5.18, the ASK2 and the ASK4 signals are classified as MASK and also, the FSK2 and the FSK4 are classified as MFSK. Meanwhile the results in Table 5.19, represent the performance evaluation for the ASK2 and the ASK4 signals, using the second ANN, at the SNR of 15 dB and 20 dB. Also, the results in Table 5.20 represent the performance evaluation for the FSK2 and the FSK4 signals, using the third ANN, at a SNR of 15 dB and 20 dB. From Tables 5.17 and 5.19 (15 dB SNR), it is clear that all the types of analogue and digital modulations have been correctly classified with more than 86.0% success rate. Excluding AM, ASK4, and FM, the success rate is $> 96.0\%$ at the SNR of 15 dB for all analogue and digital modulations. At the SNR of 20 dB (see Tables 5.18 and 5.20) all the modulation types have been classified with success rate $> 93.0\%$ except the AM (=87.6%) and ASK4 (=85.8%). Excluding AM, ASK4, and FM, the success rate is $> 97.0\%$ for all analogue and digital modulations at the SNR of 20 dB .

5.5.3   Speed-up of the training time

Similar to the AMRAs and the DMRAs, a normalisation for the datasets is used to speed up the training phase of the developed algorithm for the recognition of both analogue and digital modulations. Fig. 5.19 represents the SSE values for only 250,000 training epochs with and without normalisation. From Fig. 5.17 and Fig. 5.19, it is clear that applying the normalisation introduces SSE at 250,000 training epochs only less than the SSE for the same network structure (15-15) without normalisation and using 250,000 epochs for both training and learning. So, the normalisation increases the speed of the training phase at least twice with respect to that without normalisation.

5.6   Summary of Results & Performance Comparisons

The datasets used in this chapter are identical to those used in Chapter 2 for analogue modulations, Chapter 3 for digital modulations, and Chapter 4 for both analogue and digital modulations. Therefore direct comparisons of these two approaches can be made. Results for analogue and digital modulation recognition are summarised in Tables 5.21 and 5.22. The overall success rates of the decision-theoretic & the ANN algorithms for correct modulation recognition, based on 400 realizations for each modulated signal of interest have been measured at different SNR values. The overall success rate of the AMRAs using the DT approach and those using the ANN with single hidden layer (25 nodes), and two hidden layers (10-10) are presented in Table

5.21 at the SNR of 5 dB, 10 dB and 20 dB. Also, the overall success rate of the DMRAs using the DT and those using the ANN with single hidden layer (10 nodes), and two hidden layers (12-12) are presented in Table 5.21 at the SNR of 5 dB, 10 dB and 20 dB. Furthermore, the overall success rate of the the ADMRAs using the DT and that using the ANN (15-15) are presented in Table 5.22 at the SNR of 5 dB, 10 dB, 15 dB and 20 dB.

In Table 5.21, each number represents the average probability of the correct classification of either the twelve analogue or the six digitally modulated signals. Also, the numbers in Table 5.22 represent the average probability of correct classifications of both analogue and digital modulations. Given that the results in Chapters 2, 3, and 4 and Appendix D were comparatively very good, the ANN results are very encouraging and point towards the adoption of ANN approaches. Generally, the results obtained from the ANN approach are better than those reported by the DT approach as seen from Tables 5.21 and 5.22.

5.7   Conclusions

In the ANN algorithms for modulation recognition a lot of work has been done. Three types of artificial neural network - no hidden layer, one hidden layer, and double hidden layers ANNs - have been considered. In the AMRAs and the DMRAs, both the one hidden layer and the two hidden layers are considered as two groups of algorithms for modulation recognition using the ANN approach. In the ADMRAs, the no hidden layer and the two hidden layers are used to choose the best ADMRA structure. The same datasets used in the decision-theoretic approach are also used in the ANN approach. It is worth noting that, the training has been done using only 50 realizations at each SNR (10 dB and 20 dB for both the AMRAs and the DMRA, and 15 dB and 20 dB for the ADMRAs), while the performance is measured for the 400 realizations of each modulated signal of interest at the corresponding SNR values used in the training stage.

In the AMRAs using the ANN approach, many network structures have been tested to choose the best, that give acceptable success rate, while the others have been discarded. In the single hidden layer case it was found that the best network has 25 nodes in the hidden layer, and is achieved overall success rate 98.9% at the SNR of

10 dB. Also, using this network, all the analogue modulation types of interest have been classified with success rate > 98.0% except the USB (95.8% success rate) at the SNR of 10 dB. In the two hidden layers network, it was found that five structures have overall success rates > 99.0%. One of them - (10-10) - is considered further to measure the performance of the developed algorithm. In this case, the chosen network achieves overall success rate 99.6% at the SNR of 10 dB.

In the DMRAs using the ANN approach, many network structures have been tested to choose the best. In the single hidden layer network it was found that the best has 10 nodes in the hidden layer, and it was found that its overall success rate 97.5% at the SNR of 10 dB. In this case, all the digital modulation types of interest have been classified with success rate > 95.0% except ASK2 (=93.5%) at the SNR of 10 dB. In the two hidden layers network, it was found that five structure have success rate > 97.0%. One of them - (12-12) - is considered further to measure the performance of the developed algorithm. In this case, the chosen network achieves overall success rate 97.4% at the SNR of 10 dB. As the SNR increase to 20 dB, the algorithm does better and now all the digital modulation types of interest have been classified with success rate > 99.0%.

In the ADMRAs using the ANN approach, the best ANN structure for this kind of algorithms as shown in Fig. 5.16, comprising three networks. The first network is more complex compared to the other two networks, as most of the modulation types are classified using the first network. The first network is of the two hidden layer type while the second and the third networks are of the no hidden layers type. In this algorithm, all the modulation types of interest (analogue and digital) are classified with overall success rate > 96.0% at the SNR of 15 dB and 20 dB.

Furthermore, the reduction of the training time for all the algorithms presented is done by normalising the datasets used for the training phase. It is shown that this normalisation speeds up the training phase of the proposed ANN algorithms for modulation recognition. For example; in the AMRAs a reduction in the training time of at least twice for the single hidden layer is achieved and at least 4 times for the double hidden layer ANN algorithms. In the DMRAs, the reduction in the training time is at least 3 times for the single hidden layer and it is at least 6 times for the double hidden layer ANN algorithm. In the ADMRAs, the reduction in the training time is at least twice.

Finally the most important conclusion is that the ANN removes the need for separate threshold determination, removes the active choice of the time-ordering of threshold testing, and often offers better success rate than the decision-theoretic algorithms, introduced in Chapters 2, 3, and 4 and Appendix C.

Number of nodes in hidden layer	Average probability of correct decision
20	98.7%
25	99.4%
30	98.8%

Table 5.1: Overall performance for the single hidden layer ANN of all the tested structures *[based on 100,000 epochs for both training and learning]* for the AMRAs.

Number of nodes in the hidden layers		Average Probability of correct decisions
First	Second	
5	5	96.7%
5	7	99.2%
7	5	99.5%
7	7	99.4%
7	10	96.7%
10	7	99.6%
10	10	99.6%
12	12	99.8%

Table 5.2: Overall performance for the two hidden layers ANN of all the tested structures *[based on 100,000 training epochs]* for the AMRAs.

Number of nodes in the hidden layers		Average Probability
First	Second	of correct decisions
5	7	99.2%
7	5	97.6%
7	7	99.6%
10	7	99.7%
10	10	99.6%
12	12	99.8%

Table 5.3: Overall performance for the two hidden layers ANN of the best six structures *[based on 100,000 epochs for both training and learning]* of the AMRAs.

Simulated Type	Deduced modulation type						
	AM	DSB	VSB	LSB	USB	COM.	FM
AM	100%	-	-	-	-	-	-
DSB	-	100%	-	-	-	-	-
VSB	-	-	98.0%	-	-	2.0%	-
LSB	-	-	-	100 %	-	-	-
USB	-	-	4.2%	-	95.8%	-	-
COM.	-	-	-	-	-	99.7%	0.3%
FM	-	-	-	-	-	0.9%	99.1%

Table 5.4: Performance of the single hidden layer (25 nodes) AMRA at 10 dB SNR.

Simulated Type	Deduced modulation type						
	AM	**DSB**	**VSB**	**LSB**	**USB**	**COM.**	**FM**
AM	100%	-	-	-	-	-	-
DSB	-	100%	-	-	-	-	-
VSB	-	-	100%	-	-	-	-
LSB	-	-	-	100%	-	-	-
USB	-	-	5.7%	-	94.3%	-	-
COM.	-	0.1%	-	-	-	99.7%	0.2%
FM	-	-	-	-	-	-	100%

Table 5.5: Performance of the single hidden layer ANN (25 nodes) AMRA at 20 dB SNR.

Simulated Type	Deduced modulation type						
	AM	**DSB**	**VSB**	**LSB**	**USB**	**COM.**	**FM**
AM	100%	-	-	-	-	-	-
DSB	-	100%	-	-	-	-	-
VSB	-	-	98.0%	-	2.0%	-	-
LSB	-	-	-	98.8%	-	1.2%	-
USB	-	-	-	-	99.3%	0.7%	-
COM.	-	-	-	0.3%	-	99.1%	0.6%
FM	-	-	-	-	-	0.1%	99.9%

Table 5.6: Performance of the two hidden layers (10-10) AMRA at 10 dB SNR.

Simulated Type	Deduced modulation type						
	AM	DSB	VSB	LSB	USB	COM.	FM
AM	100%	-	-	-	-	-	-
DSB	-	100%	-	-	-	-	-
VSB	-	-	100%	-	-	-	-
LSB	-	-	-	99.5%	-	0.5%	-
USB	-	-	-	-	99.8%	0.2%	-
COM.	-	0.1%	-	0.1%	-	99.5%	0.3%
FM	-	-	-	-	-	-	100%

Table 5.7: Performance of the two hidden layers (10-10) AMRA at 20 dB SNR.

Number of nodes in hidden layer	Average probability of correct decision
5	82.3%
10	98.5%
15	96.7%

Table 5.8: Overall performance for the single hidden layer ANN of all the tested structures *[based on 150,000 epochs for both training and learning]* for the DMRAs.

Number of nodes in the hidden layers		Average Probability
First	Second	of correct decisions
5	5	94.8%
5	7	94.8%
7	5	96.6%
7	7	97.8%
7	10	95.5%
10	7	96.6%
10	10	97.2%
12	12	98.4%

Table 5.9: Overall performance for the two hidden layers ANN of all the tested structures *[based on 150,000 training epochs]* for the DMRAs.

Number of nodes in the hidden layers		Average Probability
First	Second	of correct decisions
7	5	98.1%
7	7	98.1%
10	7	97.6%
10	10	97.9%
12	12	98.6%

Table 5.10: Overall performance for the two hidden layers ANN of the best six structures *[based on 150,000 epochs for both training and learning]* for the DMRAs.

Simulated Type	Deduced modulation type					
	ASK2	ASK4	PSK2	PSK4	FSK2	FSK4
ASK2	93.5%	6.5%	-	-	-	-
ASK4	0.2%	99.8%	-	-	-	-
PSK2	-	-	100%	-	-	-
PSK4	-	-	1.2%	95.8%	3.0%	-
FSK2	-	-	-	-	95.8%	4.2%
FSK4	-	-	-	-	0.2%	99.8%

Table 5.11: Performance of the single hidden layer (10 nodes) DMRA at 10 dB SNR.

Simulated Type	Deduced modulation type					
	ASK2	ASK4	PSK2	PSK4	FSK2	FSK4
ASK2	100%	-	-	-	-	-
ASK4	0.5%	99.5%	-	-	-	-
PSK2	-	-	100%	-	-	-
PSK4	-	-	1.7%	97.8%	0.5%	-
FSK2	-	-	-	-	100%	-
FSK4	-	-	-	-	-	100%

Table 5.12: Performance of the single hidden layer (10 nodes) DMRA at 20 dB SNR.

Simulated Type	Deduced modulation type					
	ASK2	ASK4	PSK2	PSK4	FSK2	FSK4
ASK2	97.0%	3.0%	-	-	-	-
ASK4	0.2%	99.8%	-	-	-	-
PSK2	-	-	100%	-	-	-
PSK4	-	0.2%	1.0%	96.3%	2.5%	-
FSK2	-	0.2%	-	0.3%	92.5%	7.0%
FSK4	-	1.0%	0.2%	-	0.3%	98.5%

Table 5.13: Performance of the two hidden layers (12-12) DMRA at 10 dB SNR.

Simulated Type	Deduced modulation type					
	ASK2	ASK4	PSK2	PSK4	FSK2	FSK4
ASK2	100%	-	-	-	-	-
ASK4	-	100%	-	-	-	-
PSK2	-	-	99.8%	0.2%	-	-
PSK4	-	-	-	99.5%	0.5%	-
FSK2	-	-	-	-	100%	-
FSK4	-	-	-	-	-	100%

Table 5.14: Performance of the two hidden layers (12-12) DMRA at 20 dB SNR.

Number of nodes in the hidden layers		Average Probability
First	**Second**	**of correct decisions**
5	5	81.8%
10	10	89.6%
10	12	93.5%
10	15	91.5%
12	10	91.8%
12	12	95.0%
12	15	94.7%
15	10	92.5%
15	12	91.9%
15	15	93.4%
15	18	84.4%
18	15	94.1%
18	18	95.0%
20	20	84.0%

Table 5.15: Overall performance for the first network of all the tested structures *[based on 100,000 training epochs]* in the ADMRAs.

Number of nodes in the hidden layers		Average Probability
First	**Second**	**of correct decisions**
10	12	93.4%
12	12	95.0%
12	15	95.6%
15	15	95.9%
18	15	93.6%
18	18	96.2%

Table 5.16: Overall performance of the first network of the best six structures *[based on 250,000 epochs for both training and learning]* in the ADMRAs.

Simulated Type	Deduced Modulation Type										
	AM	DSB	VSB	LSB	USB	COM.	FM	MASK	PSK2	PSK4	MFSK
AM	88.5%	-	0.1%	-	-	-	-	11.4%	-	-	-
DSB	-	100%	-	-	-	-	-	-	-	-	-
VSB	-	-	100%	-	-	-	-	-	-	-	-
LSB	0.2%	-	-	99.8%	-	-	-	-	-	-	-
USB	-	-	1.0%	-	98.5%	-	-	0.2%	0.3%	-	-
Combined	-	-	-	-	-	97.4%	-	-	-	2.3%	0.3%
FM	-	-	-	-	-	-	90.1%	-	-	-	9.9%
ASK2	3.2%	-	-	-	-	-	-	96.8%	-	-	-
ASK4	13.5%	-	-	-	-	-	-	86.5%	-	-	-
PSK2	0.5%	-	-	-	-	-	-	-	99.5%	-	-
PSK4	-	-	-	-	-	0.8%	0.2%	-	1.2%	96.8%	1.0%
FSK2	-	-	-	-	-	-	1.0%	-	-	-	99.0%
FSK4	-	-	-	-	-	-	0.5%	-	-	-	99.5%

Table 5.17: Performance of the first ANN in the ADMRA (15-15) at SNR = 15 dB.

Simulated Type	Deduced Modulation Type										
	AM	DSB	VSB	LSB	USB	COM.	FM	MASK	PSK2	PSK4	MFSK
AM	87.6%	-	-	-	-	-	-	12.4%	-	-	-
DSB	-	100%	-	-	-	-	-	-	-	-	-
VSB	-	-	100%	-	-	-	-	-	-	-	-
LSB	-	-	-	100%	-	-	-	-	-	-	-
USB	-	-	1.0%	-	98.5%	-	-	0.2%	-	0.3%	-
Combined	-	-	-	-	-	97.6%	-	-	-	1.9%	0.5%
FM	-	-	-	-	-	-	93.4%	-	-	-	6.6%
ASK2	3.0%	-	-	-	-	-	-	97.0%	-	-	-
ASK4	14.2%	-	-	-	-	-	-	85.8%	-	-	-
PSK2	0.5	-	-	-	-	-	-	-	99.5%	-	-
PSK4	-	-	-	-	-	0.3%	0.2%	-	1.5%	97.5%	0.5%
FSK2	-	-	-	-	-	-	1.0%	-	-	-	99.0%
FSK4	-	-	-	-	-	-	2.5%	-	-	-	97.5%

Table 5.18: Performance of the first ANN in the ADMRA (15-15) at SNR = 20 dB.

Type	SNR = 15 dB		SNR = 20 dB	
	ASK2	ASK4	ASK2	ASK4
ASK2	97.0%	3.0%	100%	-
ASK4	-	100%	-	100%

Table 5.19: Performance of the second ANN.

Type	SNR = 15 dB		SNR = 20 dB	
	FSK2	FSK4	FSK2	FSK4
FSK2	97.5%	2.5%	100%	-
FSK4	-	100%	-	100%

Table 5.20: Performance of the third ANN.

SNR (dB)	AMRAs					DMRAs			
	DT			ANN		DT		ANN	
	I&V	II&III	IV	1-layer	2-layer	I & II	III	1-layer	2-layer
5	93.0%	90.3%	91.8%	92.3%	95.4%	55.1%	55.1%	62.2%	70.3%
10	99.4%	97.2%	98.0%	98.9%	99.6%	99.0%	99.0%	97.5%	97.4%
20	99.8%	97.7%	98.8%	99.1%	100%	99.9%	100%	99.6%	99.9%

Table 5.21: Overall success rates of the AMRAs and DMRAs, using both the DT and the ANN approaches.

SNR (dB)	ADMRAs		
	DT		ANN
	I	II&III	
5	61.3%	59.1%	60.9%
10	87.9%	82.6%	88.1%
15	94.6%	93.3%	96.3%
20	94.6%	93.2%	96.4%

Table 5.22: Overall success rates of the ADMRAs, using both the DT and the ANNs approaches.

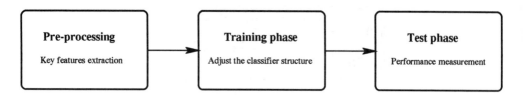

Figure 5.1: Functional blocks of the ANN algorithms.

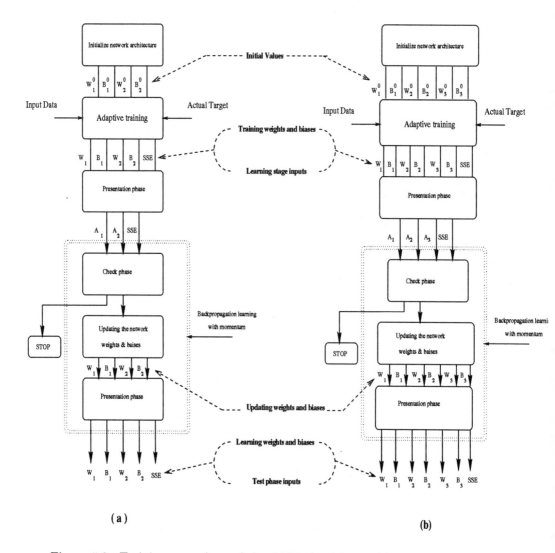

Figure 5.2: Training procedure of the ANN algorithms: (a) single hidden layer, and (b) two hidden layers.

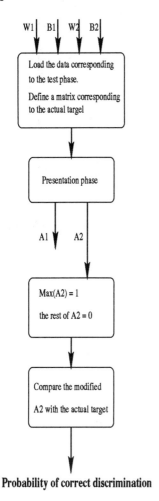

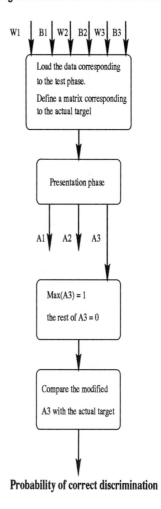

(a)

(b)

Figure 5.3: Test procedure of the ANNs algorithms: (a) single hidden layer, and (b) two hidden layers.

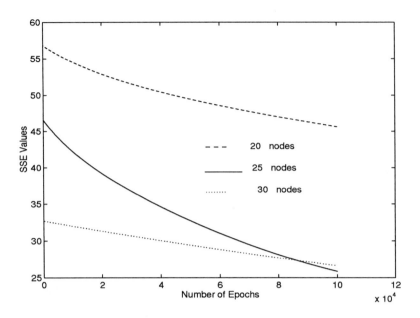

Figure 5.4: Dependence of the learning SSE for the single hidden layer ANNs [AM-RAs].

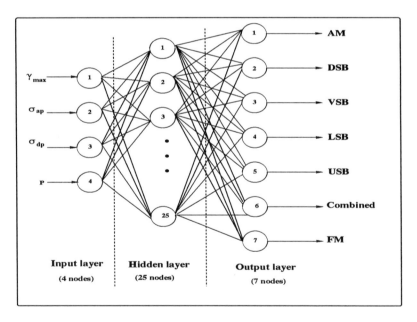

Figure 5.5: Single hidden layer ANN architecture for the AMRAs.

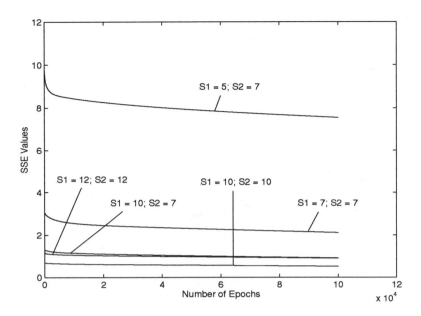

Figure 5.6: Dependence of the learning SSE for the two hidden layers ANNs [AMRAs].

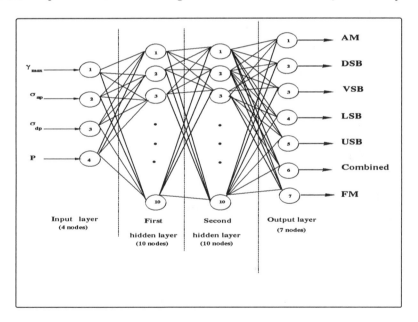

Figure 5.7: Two hidden layers ANN architecture for AMRAs.

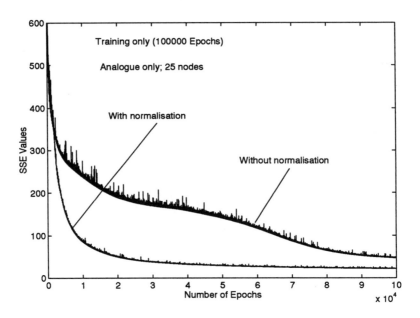

Figure 5.8: Effect of key features normalisation on the training SSE for the single hidden layer ANNs [AMRA].

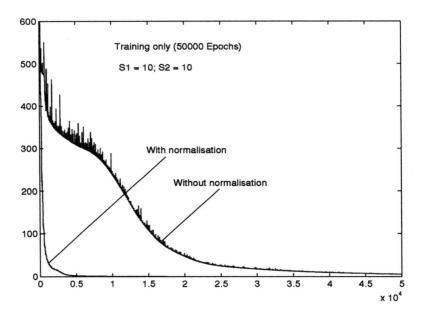

Figure 5.9: Effect of key features normalisation on the training SSE for the double hidden layer ANNs [AMRA].

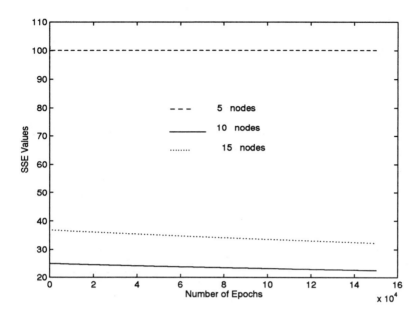

Figure 5.10: Dependence of the learning SSE for the single hidden layer ANNs [DM-RAs].

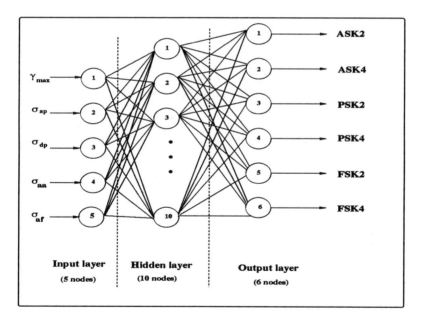

Figure 5.11: Single hidden layer ANN architecture for DMRAs.

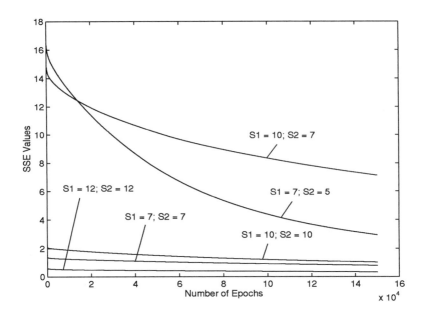

Figure 5.12: Dependence of the learning SSE for the double hidden layer ANNs [DM-RAs].

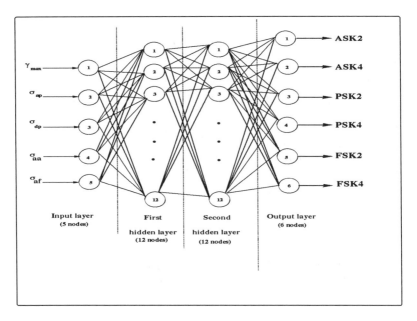

Figure 5.13: Double hidden layer ANN architecture for the DMRAs.

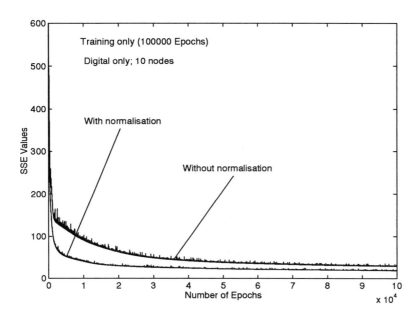

Figure 5.14: Effect of key features normalisation on the training SSE for the single hidden layer ANNs [DMRA].

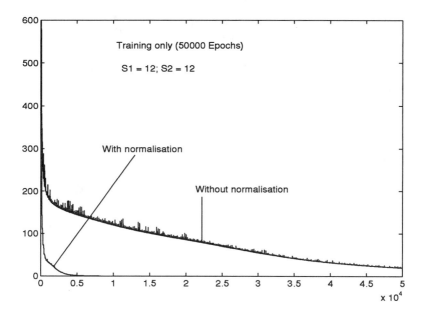

Figure 5.15: Effect of key features normalisation on the training SSE for the double hidden layer ANNs [DMRA].

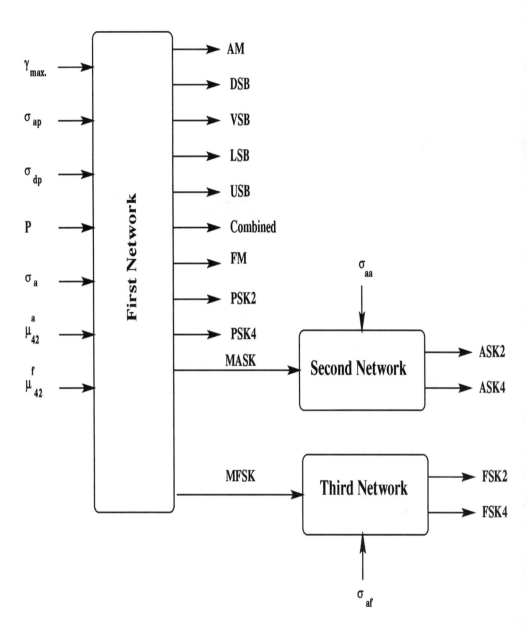

Figure 5.16: ANN structure for the ADMRAs.

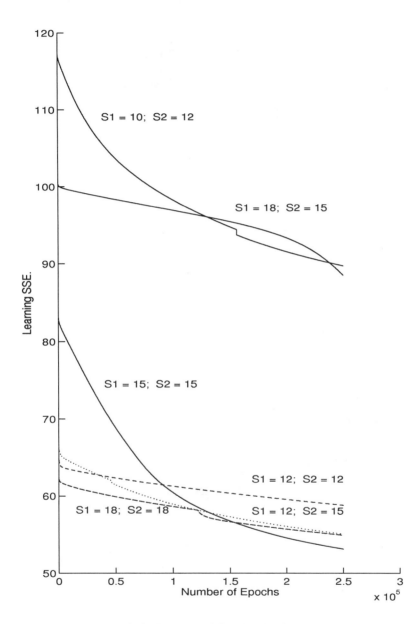

Figure 5.17: Dependence of the learning SSE for the first network of the ADMRAs.

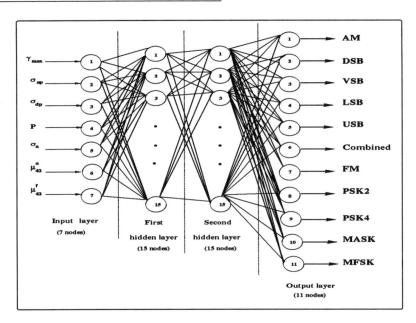

Figure 5.18: First network architecture for the ADMRAs.

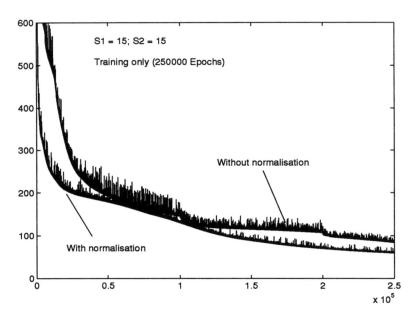

Figure 5.19: Effect of key features normalisation on the training SSE for the ADMRA.

Chapter 6

Summary and Suggestions for Future Directions

In this book, a set of fast algorithms has been introduced for on-line automatic modulation recognition of communication signals. Three groups of algorithms utilise the decision-theoretic approach in addition to another three groups of algorithms utilising the ANN approach. In each approach, one group is for the recognition of analogue modulations only, the second is for the recognition of digital modulations only, and the third is for the recognition of both analogue and digital modulations without any a priori information about the nature of the signal. A number of novel key features has been proposed to fulfil the requirement of these algorithms. Due to the simplicity of the key feature extraction (using conventional signal processing tools) and the decision rules used in the decision-theoretic approach as well as the simplicity of the network structures in the ANN approach, all the developed algorithms are seen to be suitable for on-line analysis.

Generally, the use of ANNs in the modulation recognition process removes the need for separate threshold determination, removes the active choice of the time-ordering of thresholds testing, and often offers better success rate than the decision-theoretic algorithms, introduced in Chapters 2, 3, and 4 and Appendix C. Furthermore, the performance of these modulation recognisers may be improved by applying the global procedure for modulation recognition, introduced in Section 2.2, but this requires longer time. The computing time depends mainly on the number of sample points, the number of modulation types of interest, the number of extracted key features, the nature of the extracted key features and the complexity of the classification algorithm. Anyway, the use of the global procedure for modulation recognition is postponed for the future work.

6.1 Summary by Chapters

6.1.1 Analogue modulation recognition algorithms (Chapter 2)

Five analogue modulations recognition algorithms (AMRAs) based on four key features and utilising the decision-theoretic approach are presented. These algorithms use the key features γ_{max}, σ_{ap}, σ_{dp}, and the ratio P. These key features are characterised by their simplicity as they are extracted using the conventional signal processing tools. Also, the decision rules used in these algorithms are very simple logical functions. For these reasons, these algorithms can be implemented at extremely low cost and as well they can be used for on-line analysis.

Extensive simulations of twelve analogue modulated signals have been carried out at different SNR. The mathematical formula used in the software simulations of different analogue modulation types are introduced in Section 2.3. Sample results have been presented at the SNR of 10 dB, 15 dB, and 20 dB for the developed AMRAs. Furthermore, the overall success rates for these algorithms are introduced at 10 dB, 15 dB, and 20 dB SNRs. It is observed that the threshold SNR for correct modulation recognition is about 10 dB, which is an improvement in the SNR threshold over previous results reported in [7], [10] and [19]. Also, all the presented AMRAs achieve overall success rates > 97.0%.

6.1.2 Digital modulation recognition algorithms (Chapter 3)

For digital modulations recognition, three algorithms based on five key features and utilising the decision-theoretic approach, are proposed. The key features used are γ_{max}, σ_{ap}, σ_{dp}, σ_{aa}, and σ_{af}. All these key features are extracted using the conventional signal processing tools. Also, the decision rules used in these algorithms are very simple logical functions. So, the algorithms developed are most suitable for on-line analysis.

Extensive simulations of six digitally modulated signals have been carried out at different SNR. Sample results have been presented at the SNR of 10 dB, 15 dB and 20 dB. All the modulation types of interest have been classified with success rate > 98.0% at the SNR of 10 dB. Furthermore, the overall success rates for the developed DMRAs, that utilise the decision-theoretic approach, are introduced at 10 dB, 15 dB, and 20 dB SNRs and it observed that the overall success rate is ≈ 99.0% at the SNR of 10 dB and it is ≈ 100% at the SNR of 20 dB. So, the threshold SNR for correct digital modulations recognition is about 10 dB, which is an improvement in the SNR threshold over previous results reported in [4], [14] and [18].

6.1.3 Analogue and digital modulation recognition algorithms (Chapter 4)

Based on nine key features, three algorithms for the recognition of both analogue and digital modulations are developed. The key features used are γ_{max}, σ_{ap}, σ_{dp}, P, σ_{aa}, σ_{af}, σ_a, μ_{42}^a, and μ_{42}^f. All these key features are extracted using the conventional signal processing tools. So, the developed algorithms can be implemented at extremely low cost and they can be used for on-line analysis.

In these algorithms, extensive simulations of twelve analogue and six digitally modulated signals have been carried out at different SNR. Sample results have been presented at the SNR of 15 dB and 20 dB. It is found that the threshold SNR for correct modulation recognition is about 15 dB and that they achieve overall success rate ≈ 95.0% for the ADMRA I and for the ADMRAs II & III, the overall success rate is ≈ 93.0%. So, in the algorithms developed it can be noticed that there an improvement both in the reduced SNR threshold over previous results reported in [5] and [16], and in the larger range of modulation types considered.

6.1.4 Modulation recognition using ANNs (Chapter 5)

Extensive simulations for twelve analogue and six digitally modulated signals have been carried out to measure the performance of the algorithms presented. Sample results are introduced at two SNR values (10 dB and 20 dB for both analogue only and digital only algorithms, and 15 dB and 20 dB for the recognition of both analogue and digital modulations without any a priori information).

Generally, the separation between the different analogue modulations with the use of the single hidden layer - 25 nodes - ANN has been achieved with overall success rate > 98.0% at the SNR of 10 dB. Also, the separation between the different digital modulations with the use of the single hidden layer - 10 nodes - ANN has been achieved with overall success rate > 97.0% at the SNR of 10 dB. Furthermore, the ADMRA achieves overall success rate > 96.0% at the SNR of 15 dB. Reduction of the training time for these algorithms is achieved by normalising the datasets used in training phase. It was found that the normalisation achieves a reduction in the training time of at least twice for the single hidden layer AMRA, 4 times for the two hidden layers AMRA, 3 times for the single hidden layer DMRA, 6 times for the two hidden layer DMRA, and twice for the ADMRA.

It should be noted that the results of the ANN algorithms were derived by training networks using only 1/8 of the test data, but for the DT algorithms all the test data were used to determine the optimum key features threshold values.

6.2   Suggestions for Future Directions

Finally, the authors suggest the following extensions for this work.

1. Apply the global procedure for modulation recognition, introduced in Section 2.2, which may improve the performance of the decision-theoretic algorithms.

2. Check the suitability of the DMRAs and the ADMRAs for a larger number of levels, $M > 4$ in the digital modulations.

3. Find more methods for reducing the training time of the ANN algorithm to increase the validity of this approach to be suitable for on-line analysis.

4. Apply the algorithms developed to real signals.

5. Reside the softwares developed in an EPROM to be available for real field applications.

6. Study the effect of higher order statistics and cyclostationarity on the modulation recognition problem, since it may be possible to find other key features that may be less sensitive to noise.

It is deemed that the above proposed extensions would reduce the complexity and the required threshold SNR for correct modulation recognition.

Bibliography

[1] C. B. Hafmann and A. R. Baron, " Wide-band ESM receiving systems," Microwave J., September 1980, pp. 23,24.

[2] D. Torrieri, "Principle of military communication system," Artech House, Inc., Washington, 1981.

[3] B. T. James, " Microwave receivers with electronic warfare applications," Centre Vill, Ohio, March 1986.

[4] F. F. Liedtke, "Computer simulation of an automatic classification procedure for digitally modulated communication signals with unknown parameters," Signal Processing, Vol. 6, No. 4, August 1984, pp. 311-323.

[5] T. G. Callaghan, J. L. Pery, and J. K. Tjho, "Sampling and algorithms aid modulation recognition," Microwaves RF, Vol. 24, No. 9, September 1985, pp. 117-119, 121.

[6] F. Jondral," Automatic classification of high frequency signals," Signal Processing, Vol. 9, No. 3, October 1985, pp. 177-190.

[7] P. M. Fabrizi, L. B. Lopes and G. B. Lockhart, " Receiver recognition of analogue modulation types," IERE Conference on Radio Receiver and Associated Systems, Bangor, Wales, 1986, pp. 135-140.

[8] J. Aisbett, " Automatic modulation recognition using time domain parameters," Signal Processing, Vol. 13, No. 3, October 1987, pp. 323-328.

[9] M. P. DeSimio and E. P. Glenn, "Adaptive generation of decision functions for classification of digitally modulated signals," NAECON, 1988, pp. 1010-1014.

[10] Y. T. Chan and L. G. Gadbois, "Identification of the modulation type of a signal," Signal Processing, Vol. 16, No. 2, February 1989, pp. 149-154.

[11] P. M. Petrovic, Z. B. Krsmanovic and N. K. Remenski," An automatic VHF signal classifier", Mediterranean Electro technical Conference, MELECON, 1989, pp. 385-387.

[12] A. Martin, " A signal analysis and classification strategy for implementation in an EW communication receiver," Fifth International Conf. on Radio Receiver and Associated Systems, July 1990, pp. 222-226.

[13] A. Polydoros and K. Kim, " On the detection and classification of quadrature digital modulations in broad-band noise," IEEE Transactions on Communications, Vol. 38, No. 8, August 1990, pp. 1199-1211.

[14] Z. S. Hsue and S. S. Soliman, "Automatic modulation classification using zero-crossing," IEE Proc. Part F Radar signal process, Vol. 137, No. 6, December 1990, pp. 459-464.

[15] S. D. Jovanovic, M. I. Doroslovacki, and M. V. Dragosevic, " Recognition of low modulation index AM signals in additive Gaussian noise," European Association for Signal Processing V Conference, 1990, pp. 1923-1926.

[16] L. V. Dominguez, J. M. Borrallo and J. P. Garcia, " A general approach to the automatic classification of radio communication signals," Signal Processing, Vol. 22, No. 3, March 1991, pp. 239-250.

[17] S. S. Soliman and Z. S. Hsue, " Signal classification using statistical moments," IEEE Transactions on Communications, Vol. 40, No. 5, May 1992, pp. 908-916.

[18] K. Alssaleh, K. Farrell, and R. J. Mammone, " A new method of modulation classification for digitally modulated signals," MELCOM 92, Communication, Fusing, Command, Control, and Intelligence, October 1992. pp. 30.5.1-30.5.5.

[19] P. A. J. Nagy, " Analysis of a method for classification of analogue modulated radio signals", European Association for Signal Processing VII Conference 94, Edinburgh, Scotland, September 1994, pp. 1015-1018.

[20] P. A. J. Nagy, " A modulation classifier for multi-channel systems and multi-transmitter situations", MILCOM 1994 Conference, 1994.

[21] B. F. Beidas, C. L. Weber, " Higher-order correlation- based approach to modulation classification of digitally modulated signals," IEEE Journal on Selected Areas in Communications, Vol. 13. No. 1, January 1995.

[22] C. Y. Huang and A. Polydoros, " Likelihood method for MPSK modulation classification," IEEE Transactions on Communications, Vol. 43, No. 3, March 1995, pp. 1493-1504.

[23] Y. O. Al-jalili, " Identification algorithm for upper sideband and lower sideband SSB signals," Signal Processing, Vol. 42, No. 2, March 1995, pp. 207-213.

[24] Y. Yang and S. S. Soliman, "An improved moment-based algorithm for signal classification," Signal Processing, Vol. 43, No. 3, May 1995, pp. 231-244.

[25] L. L. Scharf, " Statistical signal processing: detection, estimation, and time series analysis," Addison-Wesley Publishing Company, Inc., New York, 1991.

[26] H. L. Van Trees, " Detection, Estimation and modulation theory," Vol. III, NewYork, Wiley, 1971.

[27] K. S. Shanmugam, "Digital and analogue communication systems," John wiley and sons, Inc., 1985.

[28] L. W. Couch II, "Digital and analogue communication systems; fourth edition," Maxwell Macmillan Canada, Inc., 1993.

[29] E. E. Azzouz," Signal intelligence processing: Modulation recognition," M.Sc. thesis, Military Technical College, Cairo, Egypt, February 1992.

[30] E. E. Azzouz, A. K. Nandi, M. H. El-ayadi and E. I. Eweda, "Recognition of analogue and digital modulations," Proceedings of the Third IASTED International Conference 1994, Cairo, Egypt, December 1994, pp. 300-304.

[31] A. K. Nandi and E. E. Azzouz, "Recognition of analogue modulations," Signal Processing, Vol. 46, No. 2, October 1995, pp. 211-222.

[32] N. S. Jayant, " Digital coding of waveform principles and applications to speech and video," Prentice-Hall, Inc. Englewood Cliffs, New Jersey, 1984.

[33] E. E. Azzouz and A. K. Nandi, "Procedure for automatic recognition of analogue and digital modulations," Accepted for publication in IEE Proceedings - Communications.

[34] M. S. Bazaraa, C. M. Shetty, "Non-linear programming - Theory and Algorithms," John Wiley, Sons, January 1979.

[35] E. E. Azzouz and A. K. Nandi, "Automatic identification of digital modulations," Signal Processing, Vol. 47, No. 1, November 1995, pp. 55-69.

[36] E. E. Azzouz and A. K. Nandi, "Techniques for Baud duration estimation," Accepted for publication in EUSIPCO-96.

[37] E. E. Azzouz and A. K. Nandi, "Automatic modulation recognition - Part I," Accepted for publication in the Journal of the Franklin Institute.

[38] A. K. Nandi and E. E. Azzouz, "Algorithms for modulation recognition of communication signals," Submitted to IEEE Transactions on Communication.

[39] R. P. Lippmann, "An introduction to computing with neural nets," IEEE Acoustics, Speech and Signal Processing Magazine, Vol. 4, No. 2. April 1987, pp. 4-22.

[40] D. R. Hush and B. G. Horne, "Progress in supervised neural networks," IEEE Signal Processing Magazine, January 1993, pp. 8-39.

[41] D. H. Nguyen and B. Widrow, "Neural networks for self-learning control systems," IEEE Control System Magazine, 1990, pp. 18-23.

[42] B. Kosko, "Neural Network for Signal Processing," Prentice Hall, 1992.

[43] K. A. Marko, J. James, J. Dosdall, and J. Murphy, "Automative control system diagnostics using neural nets for rapid pattern classification of large data sets," Proceedings of the 1989 IEEE International Joint Conference on Neural Networks, 1989, pp. 13-16.

[44] N. V. Bhat, P. A. Minderman, Jr. T. McAvoy, and N. S. Wang, "Modelling chemical process systems via neural computation," IEEE Control System Magazine, April 1990, pp. 24-30.

[45] C. W. Anderson, J.A. Franklin, and R.S. Sutton, "Learning a nonlinear model of a manufacturing process using multilayer connectionist networks," Proceedings of the $5^t h$ IEEE International Symposium on Intelligent Control, 1990, pp. 404-409.

[46] S. Haykin, "Neural Networks: A comprehensive Foundation," Maxwell Macmillan Canada, Inc. 1994.

[47] A. K. Nandi and E. E. Azzouz, "Modulation recognition using artificial neural networks," Submitted to Signal Processing.

[48] E. E. Azzouz and A. K. Nandi, "Automatic modulation recognition - Part II," Accepted for publication in the Journal of the Franklin Institute.

[49] M. I. Skolnik, "Introduction to radar systems," second edition, McGraw-Hill, Inc., New York, 1980.

[50] V. B. Herbert, "Toward a unified theory of modulation part I : Phase-Envelope relationships," Proceeding IEEE, Vol. 53, No. 3, March 1966, pp. 340-352.

[51] V. B. Herbert, "Toward a unified theory of modulation part II : Zeros Relationships," Proceeding IEEE, Vol. 54, No. 5, May 1966.

Appendix A

Numerical problems associated with the evaluation of the instantaneous amplitude, phase and frequency.

All the modulation recognition algorithms introduced in this thesis require the evaluation of the instantaneous amplitude and the instantaneous phase to extract most of the key features used. There are some problems associated with the numerical computation of the instantaneous amplitude and phase of a real signal, such as: the speed of computation, the effect of noise on the weak intervals of a signal segment, the linear phase component, the phase wrapping, and the carrier frequency estimation. Also, in addition to these problems, there is a numerical problem associated with the evaluation of the instantaneous frequency, and it is concerned with the numerical differentiation. In this appendix, all these problems are considered along with suitable solutions, except for the carrier frequency estimation, that will explained in Appendix B with some simulation results.

A.1   Choice of sampling rate

From Shannon's sampling theorem [27], $f_s > 2f_{max}$, where f_s is the sampling frequency and f_{max} is the maximum frequency of the intercepted signal spectrum. This choice is to preserve the information content and to avoid the frequency aliasing problem. In practice, there might be a need for over sampling. Usually, f_s is chosen to be (4:8) f_c, where f_c is the carrier frequency of the intercepted signal. There are many reasons for the need of over sampling. These are:

- accurate estimation of f_{max} requires long signal duration and high SNR. In communication intelligence applications, it is very common to have short signal duration, moderate or weak SNR, and rough information about the carrier frequency. Thus, it is highly probable that no accurate estimate of f_{max} is available.

- if the signal is not band-limited, it has spectrum tails. The presence of spectrum tails makes the accurate estimation of f_{max} more difficult. In this case f_{max} can be defined only through a proper definition of the spectrum's practical zero.

- the carrier frequency estimation is performed either in the time-domain utilising the zero-crossing technique [14] or in the frequency-domain using the frequency centre method [49]. The first method implies over-sampling in order to ensure good estimation accuracy. The second method is badly affected by under-sampling ($f_s < 2f_{max}$) due to spectrum aliasing.

- if the instantaneous phase is calculated modulo-π, then in order to make phase unwrapping possible, it should be guaranteed that the phase change between two successive samples never exceeds $\frac{\pi}{2}$. This implies that the sampling rate, f_s, be larger than $4f_{max}$.

- the direct time-domain calculation of the analytic representation $z(t)$ of a real signal $x(t)$ implies the determination of its Hilbert transform $y(t)$ first [42]. Its values at the sampling instants i/f_s are approximately evaluated in the form of discrete convolution. However, the direct time-domain evaluation of the Hilbert transform requires over sampling.

$$y(i) = \frac{1}{f_s} \sum_{j=1; j \neq i}^{N} \frac{x(j)}{\pi(i-j)} \tag{A.1}$$

A.2   Speed of computation

The instantaneous amplitude and the instantaneous phase can be calculated using either the analytic representation, $z(t)$, or the complex envelope, $\alpha(t)$, of a real signal as [43]

$$a(t) = \mid z(t) \mid = \mid \alpha(t) \mid \tag{A.2}$$

and

$$\phi(t) = \arg\{z(t)\} = \arg\{a(t)\} + 2\pi f_c t \qquad (A.3)$$

There are two methods for computing the analytic representation of a real signal. The first method is the direct time-domain evaluation. In this case $z(t) = x(t) + j\, y(t)$; where $y(t)$ is the Hilbert transform of the real signal, $x(t)$, as defined in (1.4). The discrete time domain convolution is given by (A.1). The second method is through the use of the FFT algorithm. It comprises the following steps:

1. computation of the spectrum $X(f)$ of the real signal $x(t)$, using the FFT algorithm;

2. evaluation of the analytic signal spectrum $Z(f)$ as

$$Z(f) = 2U(f)X(f) \qquad (A.4)$$

where $U(f)$ is the unit step function defined in the frequency-domain; and

3. determination of $z(t)$ from $Z(f)$, through the second use of the FFT algorithm.

$$z(t) = IFFT\{Z(F)\} \qquad (A.5)$$

It was found that the frequency-domain calculation is much faster than the direct time-domain calculation.

A.3   Weak intervals of a signal segment

The effect of noise mainly appears on the instantaneous phase computation. In some modulation types such as AM (with high modulation depth), DSB, and MPSK the carrier frequency may be either severely suppressed or absent. Consequently, there are weak signal intervals where the instantaneous phase or frequency estimations are more sensitive to the noise. This is equivalent to receiving a signal with very low SNR. Instead of using a phase-locked loop (PLL) as a hardware solution for this problem [5], the author introduces two software solutions. Both of them start by localising the weak signal samples by simply checking the inequality $a_n(i) \leq a_t$ for every sample, where $a_n(i)$ is the normalised instantaneous amplitude sequence and a_t is a chosen threshold defined for the weak intervals of a signal segment. The first solution consists in assigning a constant phase, (e.g. $\pi/2$), to these samples. The second solution is more sophisticated and consists in extrapolating the phase in the localised weak intervals utilising the estimated values of the instantaneous phase in the non-weak signal intervals. The first solution is adequate for the AMRAs, presented in Chapter 2. Also it is adequate for every part of the DMRAs, presented in Chapter 3, except the estimation of the number of levels for MFSK signals. For the ADMRAs, the second solution is used to discriminate between the FM and MFSK signal and to estimate the number of levels of the MFSK signal. On the other hand the first solution is adequate for the other parts of the ADMRAs. The second solution (extrapolation technique) is presented by *Azzouz and Nandi* in [35].

A.4   Phase wrapping

The instantaneous phase is evaluated from the analytic representation of the the intercepted signal. Evidently, the instantaneous phase computation is sensitive to the error in the near-zero values of both the real and the imaginary parts of a real signal analytic representation. Computation errors might reverse the polarity of these small values and thus introduce jump discontinuities in the instantaneous phase. Therefore, it is necessary to define a practical zero for both the real and the imaginary parts of the intercepted signal analytic representation.

The modulo-2π computation of the phase sequence $\{\phi(i)\}$, causes phase wrapping when the true value of the phase goes outside the interval $[0, 2\pi]$. The linear phase component due to carrier frequency is the prime contributor to phase wrapping. Thus, it is necessary to apply one of the phase unwrapping algorithms on the modulo-2π phase sequence. The chosen phase unwrapping algorithm [50] and [51] consists in adding a correction phase sequence $\{C_k(i)\}$ to the modulo-2π instantaneous phase sequence $\{\phi(i)\}$ as follows:

$$C_k(i) = \begin{cases} C_k(i-1) - 2\pi & \text{if } \phi(i+1) - \phi(i) > \pi \\ C_k(i-1) + 2\pi & \text{if } \phi(i) - \phi(i+1) > \pi \\ C_k(i-1) & \text{elsewhere} \end{cases} \tag{A.6}$$

Such that $C_k(0) = 0$. Thus, the unwrapped phase sequence $\{\phi_{uw}(i)\}$ is given by

$$\phi_{uw}(i) = \phi(i) + C_k(i) \tag{A.7}$$

A.5   Linear-phase component

The unwrapped phase sequence has a linear phase component. The main contributor for the linear phase component is the carrier frequency. Also, over a given modulated signal segment both the noise and the modulating signal may make a small contribution to the linear phase component. It is worth noting that the unwrapped phase sequence may differ from the true phase sequence by a constant. If the carrier frequency, f_c, is accurately known, the non-linear phase component can be estimated as

$$\phi_{NL}(i) = \phi_{uw}(i) - \frac{2\pi f_c i}{f_s} \tag{A.8}$$

If the carrier frequency is unknown or if it is desired to remove the residual linear phase component, the least-squares algorithm [25] can be used as follows:

Let $\left(C_1 i + C_2\right)$ represent the unknown linear phase component. Minimising the sum of squares, $\Gamma = \sum_{i=1}^{N_s} \left[\phi_{uw}(i) - C_1 i - C_2\right]^2$, the values of C_1 and C_2 are found.

A.6   Numerical derivative

In addition to the above problems associated with the evaluation of the instantaneous amplitude, phase and frequency, a problem appears only in the calculation of the instantaneous frequency, $f(t)$, which is defined as the derivative of the instantaneous phase, $\phi(t)$, as

$$f(t) = \frac{1}{2\pi} \frac{d\phi(t)}{dt}. \tag{A.9}$$

The numerical differentiation can be achieved in two ways. The first is the well known numerical difference method and it is expressed as

$$y(t) = \frac{x(t + T_s) - x(t)}{T_s} \tag{A.10}$$

where $y(t)$ is the derivative of $x(t)$ and T_s is the sampling interval ($T_s = \frac{1}{f_s}$), where f_s is the sampling rate. This method is mainly affected by the choice of sampling rate.

The second method is achieved in the frequency domain using the Fourier transform properties [26] as follows

$$y(t) = IFFT \{-j2\pi f X(f)\} \tag{A.11}$$

where $X(f)$ is the Fourier transform of $x(t)$. It is worth noting that the second method gives more smoothing in the calculation than the first method.

Appendix B

Carrier frequency estimation

In the developed algorithms for analogue and digital modulations recognition, most of the key features are derived from the the instantaneous amplitude, phase and frequency. The evaluation of the instantaneous phase and frequency requires the exact value of the carrier frequency. Also, the ratio P, that is used to measure the spectrum symmetry around the carrier frequency requires the exact value of the carrier frequency. Furthermore, due to the difficulty of estimating the maximum frequency of an RF signal, it is better to choose the sampling rate relative to the carrier frequency. For these reasons, it is necessary to know the exact value of the carrier frequency before the modulation recognition process. Thus, the author introduces a modification to the well known zero-crossing technique for carrier frequency estimation in addition to a comparison among three methods for carrier frequency estimation. Two of these three methods are in the time-domain, while the third is in the frequency-domain. So, if the carrier frequency is unknown, it can be estimated either in the frequency domain (using the frequency-centred method or using the periodogram), or in the time-domain (using the zero-crossing technique).

B.1 Frequency-domain estimation

The carrier frequency of an RF signal can be estimated in the frequency domain, using either the periodogram or the frequency-centred method.

1. The periodogram can be used for carrier frequency estimation if and only if the carrier component exists. In this method, the carrier is estimated as the peak of the periodogram. This method is based on the maximum likelihood

estimator, in which the frequency is picked as the location of the largest peak in the periodogram. In our case there are some types in which the carrier frequency is severely suppressed or absent such as SSB (LSB and USB), VSB, DSB, MPSK, and AM with high modulation depth. So, this method may be good for the signals with large carrier component such as AM with low modulation depth and MASK signals.

2. The carrier frequency can be estimated using the frequency-centred method [49] as

$$f_c = \frac{\sum_{i=1}^{N/2} i \mid Z(i) \mid^2}{\sum_{i=1}^{N/2} \mid Z(i) \mid^2},$$ (B.1)

where $\{Z(i)\}$ is the spectrum sequence of the analytic signal associated with a real signal. This method is a good estimator for symmetric signals and it is bad for asymmetric signals.

B.2 Time-domain estimation

The estimation in the time-domain is based on the zero-crossings of the RF signal. So, the carrier frequency can be estimated [14] as follows

$$f_c = \frac{M_z - 1}{2\sum_{i=1}^{M_z-1} y(i)}$$ (B.2)

where M_z is the number of zero-crossings in a signal and $\{y(i)\}$ is the zeros-crossing difference sequence, and is defined by

$$y(i) = x(i+1) - x(i); \quad i = 1, 2, ..., M_z - 1$$ (B.3)

This method is very sensitive to the noise, especially in the weak intervals of a signal. So, the author modified this method by considering the zero-crossing sequence in the non-weak intervals of a signal only. Simulation results show that the carrier frequency estimation using the modified method is error-free for most of the modulation types of interest at the the SNR of 5 dB.

B.3   Simulation results

Consequently, one of three methods for carrier frequency estimation should applied before modulation recognition process. These methods are: 1) using the frequency-centred method, 2) using all the zero-crossings of a signal and 3) using the zero-crossings of a signal in the non-weak intervals only. The performance evaluation of these three methods, derived from 100 realizations for each modulation type of interest is measured by observing the mean and standard deviation in each method for each modulation type and at different SNR (5 dB, 10 dB, 15 dB and 20 dB). It is worth noting that, all the modulation types of interest have been simulated (see Chapters 2 and 3) with carrier frequency of 150 kHz. The measured values, using the aforementioned three methods, for the 100 realizations are shown in the Tables B.1-B.4. It is clear that method III (modified zero-crossings) is the best of the three methods.

Type	Method I	Method II	Method III	Type	Method I	Method II	Method III
AM	150.4±0.1	150.2±0.3	150.0±0.0	ASK2	150.5±0.2	152.1±1.1	150.0±0.0
DSB	150.4±0.3	150.9±1.1	150.0±0.0	ASK4	150.9±0.1	151.4±0.8	150.0±0.0
VSB	151.2±0.2	150.6±0.4	150.0±0.0	PSK2	151.4±0.2	151.2±0.7	150.0±0.0
LSB	146.8±0.5	146.8±0.8	150.0±0.0	PSK4	151.4±0.4	150.9±0.9	150.0±0.0
USB	153.6±0.5	153.5±0.8	150.0±0.0	FSK2	150.5±4.1	151.1±5.2	149.5±5.0
COM.	151.6±1.8	151.3±1.4	149.8±2.0	FSK4	151.6±3.3	151.1±3.9	149.1±2.7
FM	151.0±0.9	150.6±0.8	150.0±2.3				

Table B.1: Carrier frequency estimation at SNR = 5 dB [based on 100 realizations].

Type	Method I	Method II	Method III	Type	Method I	Method II	Method III
AM	150.2±0.1	150.0±0.1	150.0±0.0	ASK2	150.4±0.1	151.1±0.7	150.0±0.0
DSB	150.2±0.2	150.6±1.1	150.0±0.0	ASK4	150.4±0.1	150.4±0.5	150.0±0.0
VSB	150.7±0.2	150.1±0.2	150.0±0.0	PSK2	150.7±0.2	151.2±0.7	150.0±0.0
LSB	146.9±0.5	146.9±0.8	150.0±0.0	PSK4	150.6±0.4	150.4±0.9	150.0±0.0
USB	153.3±0.5	153.2±0.7	150.0±0.0	FSK2	150.1±4.8	150.3±5.3	149.5±5.0
COM.	150.1±2.1	150.2±0.4	149.8±2.0	FSK4	150.1±3.8	150.3±4.1	149.1±2.7
FM	150.3±0.6	150.0±0.0	150.0±2.3				

Table B.2: Carrier frequency estimation at SNR = 10 dB [based on 100 realizations].

Type	Method I	Method II	Method III	Type	Method I	Method II	Method III
AM	150.1±0.0	150.0±0.0	150.0±0.0	ASK2	150.2±0.1	150.1±0.2	150.0±0.0
DSB	150.1±0.1	150.5±1.0	150.0±0.0	ASK4	150.2±0.1	150.0±0.1	150.0±0.0
VSB	150.6±0.2	150.0±0.1	150.0±0.0	PSK2	150.4±0.1	151.6±0.7	150.0±0.0
LSB	146.9±0.5	147.0±0.8	150.0±0.0	PSK4	150.2±0.4	150.3±0.8	150.0±0.0
USB	153.1±0.5	153.1±0.8	150.0±0.0	FSK2	150.0±5.1	150.2±5.3	150.0±0.0
COM.	150.2±2.2	150.0±0.1	150.0±0.0	FSK4	150.3±4.0	150.3±4.1	150.0±0.0
FM	150.1±0.4	150.0±0.0	150.0±0.0				

Table B.3: Carrier frequency estimation at SNR = 15 dB [based on 100 realizations].

Type	Method I	Method II	Method III	Type	Method I	Method II	Method III
AM	150.0±0.0	150.0±0.0	150.0±0.0	ASK2	150.1±0.0	150.0±0.0	150.0±0.0
DSB	150.0±0.1	150.5±1.0	150.0±0.0	ASK4	150.1±0.0	150.0±0.0	150.0±0.0
VSB	150.5±0.2	150.0±0.0	150.0±0.0	PSK2	150.3±0.1	152.3±0.8	150.0±0.0
LSB	146.9±0.5	147.0±0.8	150.0±0.0	PSK4	150.1±0.3	150.2±0.9	150.0±0.0
USB	153.1±0.5	153.1±0.8	150.0±0.0	FSK2	150.0±5.2	150.3±5.3	150.0±0.0
COM.	150.0±2.2	150.0±0.1	150.0±0.0	FSK4	150.2±4.1	150.3±4.1	150.0±0.0
FM	150.0±0.2	150.0±0.1	150.0±0.0				

Table B.4: Carrier frequency estimation at SNR = 20 dB [based on 100 realizations].

Appendix C

Alternative Algorithms for Modulation Recognition

In this appendix a set of alternative decision flows for the decision-theoretic algorithms, presented in Chapters 2, 3, and 4, is presented. In these algorithms, the modulation recognition process comprises the same steps that are presented in the corresponding algorithms in Chapters 2, 3, and 4. So, any modulation recogniser comprises two main steps: 1) pre-processing, and 2) modulation classification. It is worth noting that all these algorithms use the same key features that are used in the corresponding algorithm introduced in Chapters 2, 3 and 4, and they utilise the decision-theoretic approach.

C.1 Analogue modulation recognition algorithms

Based on the same four key features used in the proposed AMRA I, four alternative algorithms for the recognition of analogue modulations are considered here in this appendix. In these algorithms, the choice of γ_{max}, σ_{ap}, σ_{dp}, and P as key features for the alternative algorithms is based on some facts similar to that mentioned in the AMRA I (see Chapter 2). Anyway, the chosen optimum values for the key features thresholds - $t(\gamma_{max})$, $t(\sigma_{ap})$, $t(\sigma_{dp})$, and $t(P)$ - and the corresponding average probability of correct decisions (based on the 400 realizations at the SNR of 10 dB and 20 dB and for the twelve analogue modulated signals) are shown in Table C.1. Furthermore, samples results at the three SNR values - 10 dB, 15 dB, and 20 dB - are presented for these algorithms.

A detailed pictorial representation for each of the AMRAs II, III, IV and V is shown in Figs. C.1 - C.4 respectively. Based on the optimum key feature threshold values introduced in Table C.1, samples results at three SNR values - 10 dB, 15 dB, and 20 dB - are presented. It is found that the AMRA V has the same success rates for different types of analogue modulations of interest as the AMRA I. Also, the performance figures of the AMRA II and AMRA III are exactly the same. Sample results for the AMRAs II and III are presented at the SNR of 10 dB, 15 dB, and 20 dB in Tables C.2, C.3 and C.4 respectively. Also, the performance evaluation of the AMRA IV is presented at the SNR of 10 dB, 15 dB, and 20 dB in Tables C.5, C.6 and C.7 respectively.

C.2   Digital modulation recognition algorithms

Based on the same five key features used in the proposed DMRA I, two alternative algorithms for digital modulation recognition are presented. A detailed pictorial representation for each of the DMRAs II and III is shown in Fig. C.5 and Fig. C.6 respectively in the form of flowcharts.

In these algorithms, the choice of γ_{max}, σ_{ap}, σ_{dp}, σ_{aa}, and σ_{af} as key features for the alternative algorithms is based on some facts similar to those mentioned in the DMRA I (see Chapter 3). The chosen optimum values for the key features thresholds - $t(\gamma_{max})$, $t(\sigma_{ap})$, $t(\sigma_{dp})$, $t(\sigma_{aa})$, and $t(\sigma_{af})$ - and the corresponding average probability of correct decisions (based on the 400 realizations at the SNR of 10 dB and 20 dB and for the six digitally modulated signals) are given in Table C.8. Furthermore, the performance evaluations for the three algorithms are exactly the same except for the DMRA III there is a little difference at the SNR of 20 dB. The results at the SNR of 20 dB for the DMRA III are shown in Table C.9.

C.3   Analogue & digital modulation recognition algorithms

Based on the same nine key features used in the proposed ADMRA I, two more algorithms for both analogue and digital modulations recognition are presented. In these algorithms, the choice of the key features is based on the some facts similar to those mentioned in the ADMRA I (see Chapter 4). The values of the optimum key features thresholds for the ADMRAs II and III are shown in Table C.10. A detailed pictorial representation for each of the ADMRAs II and III is shown in Figs. C.7 and C.8 respectively.

The performance evaluation of the proposed algorithms is introduced for twelve analogue as well as six digitally modulated signals, and it is derived from 400 realizations for each of these signals at the SNR of 15 dB and 20 dB. The optimum values for the key features thresholds - $t(\gamma_{\max})$, $t(\sigma_{ap})$, $t(\sigma_{dp})$, $t(P)$, $t(\sigma_{aa})$, $t(\sigma_{af})$, $t(\sigma_a)$, $t(\mu_{42}^a)$, and $t(\mu_{42}^f)$ - corresponding to the ADMRA II and ADMRA III are presented in Table C.10. It is found that the performance evaluation of the ADMRA II and ADMRA III are exactly the same. The performance results are summarised in Tables C.11 and C.12 for the SNR of 15 dB and 20 dB respectively. The results represent the discrimination between the ASK2 and the ASK4 signals, and that represent the discrimination between the FSK2 and the FSK4 signals at the SNR of 15 dB and 20 dB are similar to that presented in Tables 4.5 and 4.6 for the ADMRA I.

Key features Thresholds	II		III		IV		V		Notes
	Value	P_{av}	Value	P_{av}	Value	P_{av}	Value	P_{av}	
$t(\gamma_{max})$	[5.5-6]	100%	[5.5-6]	100%	[5.5-6]	100%	[5.5-6]	100%	
$t(\sigma_{ap})$	$\pi/7$	97.0%	$\pi/7$	99.9%	$[\pi/6.5 - \pi/3]$	100%	$[\pi/6.5 - \pi/3]$	100%	DSB
					$\pi/7$	94.5%			VSB
$t(\sigma_{dp})$	$[\pi/5.5-\pi/3]$	100%	$[\pi/8 - \pi/3]$	100%	$[\pi/8 - \pi/3]$	100%	$\pi/6$	99.5%	
$t(P)$	[0.5-0.99]	100%	[0.5-0.99]	100%	0.6	100%	[0.5-0.99]	100%	SSB
	0.6	100%	0.6	100%	0.6	100%	0.6	100%	VSB

Table C.1: Optimum key features threshold values for the AMRAs - II, III, IV & V.

Simulated	Deduced Modulation Type						
Modulation Type	AM	DSB	VSB	LSB	USB	COM.	FM
AM	100%	-	-	-	-	-	-
DSB	-	99.5%	-	-	-	0.5%	-
VSB	-	-	92.8%	-	7.2%	-	-
LSB	-	0.8%	5.2%	94.0%	-	-	-
USB	-	0.5%	5.5 %	-	94.0%	-	-
COM.	-	-	-	-	-	100%	-
FM	-	-	-	-	-	-	100%

Table C.2: Confusion matrix for the AMRAs II & III [based on 400 realizations] at SNR = 10 dB

Simulated	Deduced Modulation Type						
Modulation Type	AM	DSB	VSB	LSB	USB	COM.	FM
AM	100%	-	-	-	-	-	-
DSB	-	100%	-	-	-	-	-
VSB	-	-	96.8%	-	3.2%	-	-
LSB	-	0.5%	6.5%	93.0%	-	-	-
USB	-	0.7%	6.3%	-	93.0%	-	-
COM.	-	-	-	-	-	100%	-
FM	-	-	-	-	-	-	100%

Table C.3: Confusion matrix for the AMRAs II & III [based on 400 realizations] at SNR = 15 dB

Simulated	Deduced Modulation Type						
Modulation Type	AM	DSB	VSB	LSB	USB	COM.	FM
AM	100%	-	-	-	-	-	-
DSB	-	100%	-	-	-	-	-
VSB	-	-	99.5%	-	0.5%	-	-
LSB	-	0.3%	7.2%	92.5%	-	-	-
USB	-	1.2%	7.0%	-	91.8%	-	-
COM.	-	-	-	-	-	100%	-
FM	-	-	-	-	-	-	100%

Table C.4: Confusion matrix for the AMRAs II& III [based on 400 realizations] at SNR = 20 dB

Simulated	Deduced Modulation Type						
Modulation Type	AM	DSB	VSB	LSB	USB	COM.	FM
AM	100%	-	-	-	-	-	-
DSB	-	99.5%	-	-	-	0.5%	-
VSB	-	-	92.8%	-	7.2%	-	-
LSB	-	-	-	100%	-	-	-
USB	-	-	6.0 %	-	94.0%	-	-
COM.	-	-	-	-	-	100%	-
FM	-	-	-	-	-	-	100%

Table C.5: Confusion matrix for the AMRA IV [based on 400 realizations] at SNR = 10 dB

Simulated	Deduced Modulation Type						
Modulation Type	AM	DSB	VSB	LSB	USB	COM.	FM
AM	100%	-	-	-	-	-	-
DSB	-	100%	-	-	-	-	-
VSB	-	-	96.8%	-	3.2%	-	-
LSB	-	-	-	100%	-	-	-
USB	-	-	7.0%	-	93.0%	-	-
COM.	-	-	-	-	-	100 %	-
FM	-	-	-	-	-	-	100 %

Table C.6: Confusion matrix for the AMRA IV [based on 400 realizations] at SNR = 15 dB

Simulated	Deduced Modulation Type						
Modulation Type	AM	DSB	VSB	LSB	USB	COM.	FM
AM	100%	-	-	-	-	-	-
DSB	-	100%	-	-	-	-	-
VSB	-	-	99.5%	-	0.5%	-	-
LSB	-	-	-	100%	-	-	-
USB	-	-	8.2%	-	91.8%	-	-
COM.	-	-	-	-	-	100%	-
FM	-	-	-	-	-	-	100%

Table C.7: Confusion matrix for the AMRA IV [based on 400 realizations] at SNR = 20 dB

Key Features	DMRA II		DMRA III		Notes
Thresholds	Value	$P_{av}(x_{opt})$	Value	$P_{av}(x_{opt})$	
$t(\gamma_{max})$	4	99.5%	4	99.5%	
$t(\sigma_{ap})$	$\pi/5.5$	100%	$\pi/5.5$	99.9%	
$t(\sigma_{dp})$	$[\pi/6.5 - \pi/2.5]$	100%	$[\pi/6.5 - \pi/2.5]$	100%	
$t(\sigma_{aa})$	0.25	99.6%	0.25	99.6%	
$t(\sigma_{af})$	0.40	99.8%	0.40	99.8%	

Table C.8: Optimum key features threshold values for the DMRAs II & III

Simulated	Deduced Modulation Type					
Modulation Type	ASK2	ASK4	PSK2	PSK4	FSK2	FSK4
ASK2	100%	-	-	-	-	-
ASK4	-	100%	-	-	-	-
PSK2	-	-	100%	-	-	-
PSK4	-	-	-	99.8%	0.2%	-
FSK2	-	-	-	-	100%	-
FSK4	-	-	-	-	-	100%

Table C.9: Confusion matrix for the DMRA III [based on 400 realizations] at SNR = 20 dB

Key Feature	ADMRA II		ADMRA III		Notes
Thresholds	Value	$P_{av}(x_{opt})$	Value	$P_{av}(x_{opt})$	
$t(\gamma_{max})$	[2-2.5]	100%	[2-2.5]	100%	
$t(\sigma_{ap})$	$\pi/7$	98.3%	$\pi/7$	98.5%	
$t(\sigma_{dp})$	$[\pi/7 - \pi/3]$	100%	$[\pi/6 - \pi/3]$	100%	
$t(P)$	[0.6-0.99]	100%	[0.6-0.99]	100%	SSB
	[0.5-0.7]	100%	[0.5-0.7]	100%	VSB
$t(\sigma_a)$	[0.125-0.4]	100%	[0.125-0.4]	100%	PSK2
	[0.125-0.15]	99.7%	[0.125-0.15]	99.7%	PSK4
$t(\mu_{42}^a)$	2.15	87.3%	2.15	87.3%	
$t(\mu_{42}^f)$	2.03	90.0%	2.03	90.0%	
$t(\sigma_{aa})$	0.25	99.5%	0.25	99.5%	
$t(\sigma_{af})$	0.40	99.8%	0.40	99.8%	

Table C.10: Optimum key features threshold values for the ADMRAs II & III

Simulated Types	Deduced Modulation Type										
	AM	DSB	VSB	LSB	USB	COM.	FM	MASK	PSK2	PSK4	MFSK
AM	88.8%	-	-	-	-	-	-	11.2%	-	-	-
DSB	-	100%	-	-	-	-	-	-	-	-	-
VSB	-	-	96.8%	-	3.2%	-	-	-	-	-	-
LSB	-	-	7.0%	93.0%	-	-	-	-	-	-	-
USB	-	-	7.0%	-	93.0%	-	-	-	-	-	-
Combined	-	-	-	-	-	100%	-	-	-	-	-
FM	-	-	-	-	-	-	90.0%	-	-	-	10.0%
ASK2	4.7%	-	-	-	-	-	-	95.3%	-	-	-
ASK4	22.7%	-	-	-	-	-	-	77.3%	-	-	-
PSK2	-	-	-	-	-	-	-	-	98.8%	1.2%	-
PSK4	-	-	-	-	-	0.2%	-	-	-	99.8%	-
FSK2	-	-	-	-	-	-	8.0%	-	-	-	92.0%
FSK4	-	-	-	-	-	-	12.0%	-	-	-	88.0%

Table C.11: Performance of the ADMRAs II & III at SNR = 15 dB

Simulated	Deduced Modulation Type										
Types	AM	DSB	VSB	LSB	USB	COM.	FM	MASK	PSK2	PSK4	MFSK
AM	86.1%	-	-	-	-	-	-	13.9%	-	-	-
DSB	-	100%	-	-	-	-	-	-	-	-	-
VSB	-	-	99.5%	-	0.5%	-	-	-	-	-	-
LSB	-	-	7.5%	92.5%	-	-	-	-	-	-	-
USB	-	-	8.2%	-	91.8%	-	-	-	-	-	-
Combined	-	-	-	-	-	98.8%	-	-	-	1.2%	-
FM	-	-	-	-	-	-	90.0%	-	-	-	10.0%
ASK2	4.0%	-	-	-	-	-	-	96.0%	-	-	-
ASK4	19.7%	-	-	-	-	-	-	80.3%	-	-	-
PSK2	-	-	-	-	-	-	-	-	96.3%	3.7%	-
PSK4	-	-	-	-	-	-	-	-	-	100%	-
FSK2	-	-	-	-	-	-	8.0%	-	-	-	92.0%
FSK4	-	-	-	-	-	-	12.0%	-	-	-	88.0%

Table C.12: Performance of the ADMRAs II & III at SNR = 20 dB

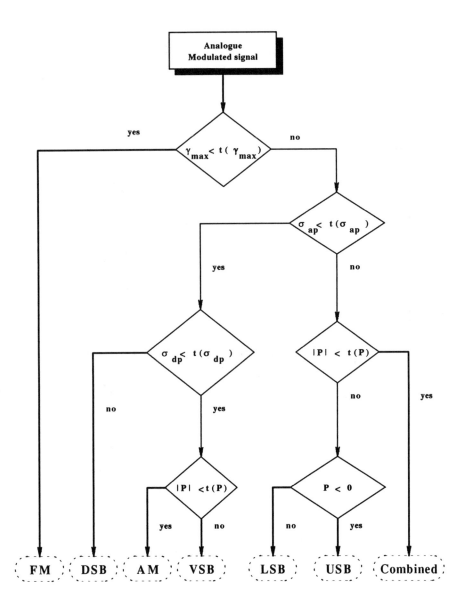

Figure C.1: Functional flowchart for the AMRA II.

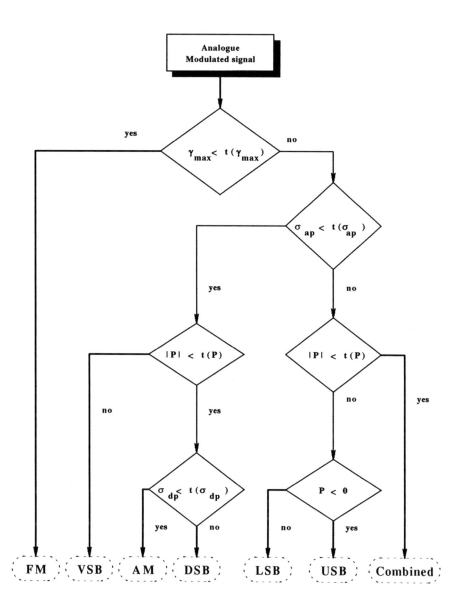

Figure C.2: Functional flowchart for the AMRA III.

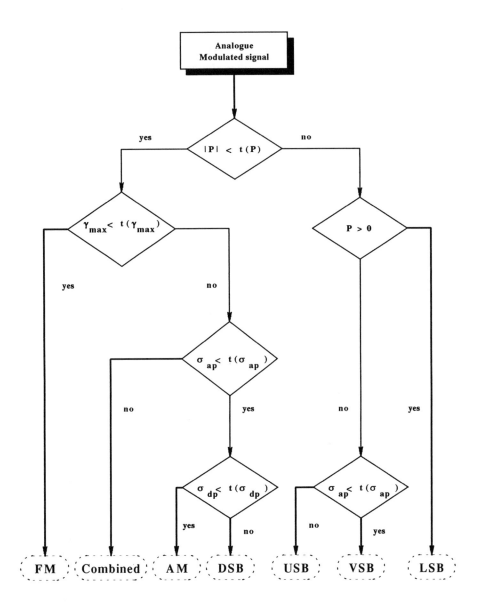

Figure C.3: Functional flowchart for the AMRA IV.

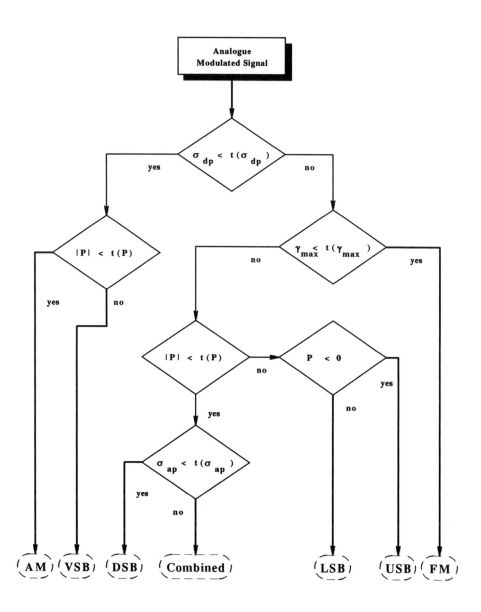

Figure C.4: Functional flowchart for the AMRA V.

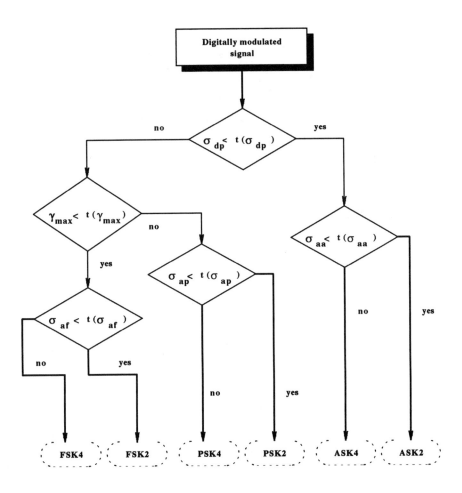

Figure C.5: Functional flowchart for the DMRA II.

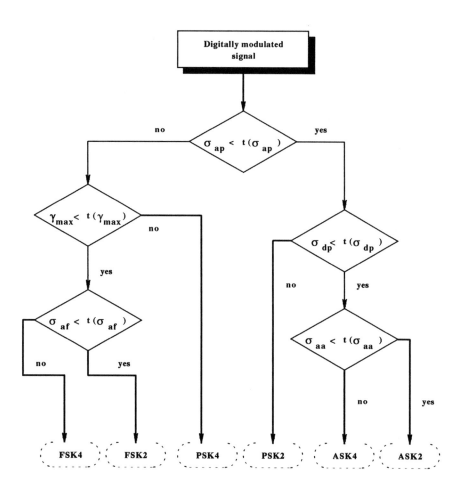

Figure C.6: Functional flowchart for the DMRA III.

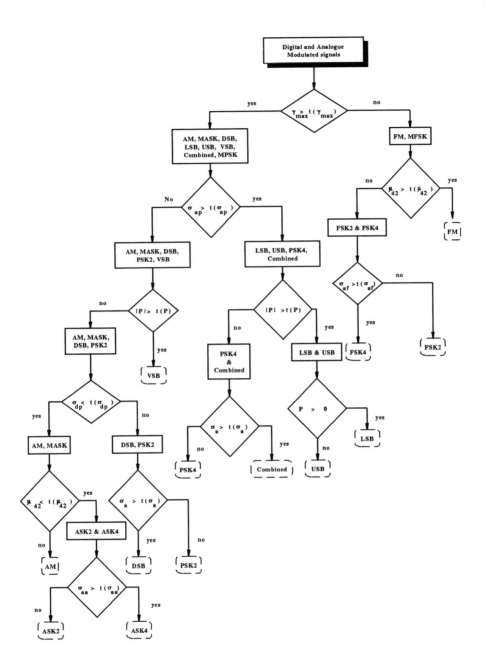

Figure C.7: Functional flowchart for the ADMRA II.

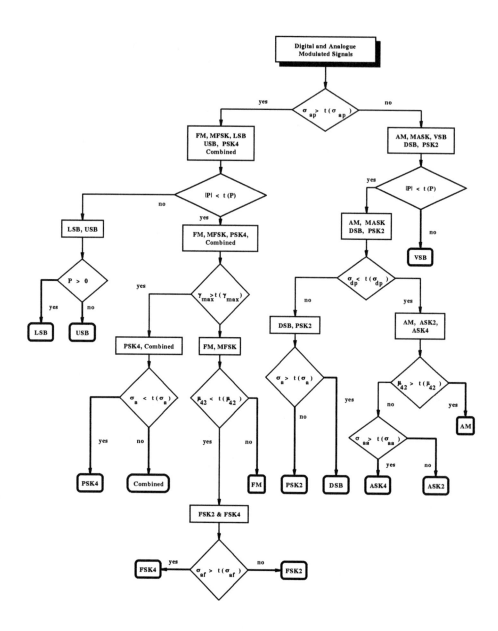

Figure C.8: Functional flowchart for the ADMRA III.

Index